COMPUTER PROGRAMS FOR MAPPING

developed by the
Experimental Cartography Unit
(to September 1970)

by T. Lang Ph D

The Architectural Press

ISBN 0 85139 128 1

This work has been supported by the
Natural Environment Research Council

The actual tapes for running these programs can be obtained on application to the Experimental Cartography Unit, and with the collaboration of the Natural Environment Research Council a price will be fixed and any special conditions laid down.

TABLE OF CONTENTS

1. Introduction

The chief function of this document is to describe the facilities which the Experimental Cartography Unit can - as at 1 Sept 1970 - make available to those interested in the applications of automated cartography. To this end, the largest part of the document, section 6, is devoted to a detailed series of instructions describing exactly how to operate the computer programs which the Unit has developed. These instructions are those which have been developed to produce maps. Most of the programs are written in FORTRAN (the remaining few being in Macro-9, the DEC assembler language), and are at present running on the PDP9 owned by the ECU. The fact that FORTRAN is a universal high level language will alleviate many of the problems associated with transferring programs to another computer, although it is fair to say that modifications may be necessary to some of the routines which access magnetic tapes and graphical devices.

Section 5 is a brief guide to the more detailed instructions, summarising what the various programs do, and how they are inter-related. A potential user who is unfamiliar with what facilities the ECU offers should first scan this section to see if there is anything applicable to his particular needs. Section 4 describes how cartographic data is stored in computer accessible form. Most importantly it specifies a general format for the storage of data on magnetic tape. Potential users of the ECU system will be strongly encouraged to adopt this format should they wish to present data for processing. Sections 2 and 3 describe how the Unit is administered and the hardware with which it is equipped.

The Unit has a continuing plan of experiment and development. Indeed, the one thing which is certain is that by the time it is published this document will be out of date. The reader is therefore cautioned against treating it as final and authoritative. It is unlikely that many facilities will be withdrawn; it is certain that very many more facilities will be added. A secondary aim of this document is to act as a historic record of the state of development reached by the Unit in September 1970. The Unit has outgrown the laboratory space available to it in Oxford, and is about to take up new premises in London, where it will be re-united with its parent body, the Royal College of Art, and will also start new links with the Imperial College of Science and Technology.

Finally, this document may also help any who wish to develop automated cartographic techniques of their own, by illustrating just what features and facilities experience has shown to be desirable.

2. Organization of the E.C.U.

The Experimental Cartography Unit was formed in 1968 under the inspiration and leadership of Mr D.P. Bickmore, formerly head of the cartographic section of the Clarendon Press, Oxford. The Unit is financed by the Natural Environment Research Council. The general policy of the Unit is guided by a Steering Committee made up of representatives of the various constituent bodies of N.E.R.C. which are actively interested in automated mapping, with, in addition, representatives of the Royal College of Art, Imperial College, the Ordnance Survey, and some computer specialists. The terms of reference of the Unit are "to plan a programme of research and development into methods of presenting data in a cartographic form using a computer technique, with emphasis, in the initial stages, on the particular needs of the N.E.R.C."

Administration for the Unit is provided by the Royal College of Art. As a significant proportion of the Unit's work is in the graphic design of computer produced maps, this relationship is not so unusual as might appear at first sight.

For convenience of operation the Unit is divided into four sections:

a) Data Capture: concerned with investigating new techniques of collecting data in forms suitable for computer processing.

b) Computer Processing: development of programs to run both on the Unit's own computer and also on larger external machines. As this document is based on a series of computer program operating instructions, it may give the false impression that this is by far the largest section. In fact, many of the programs and facilities which will be described are the result of the work or suggestions of the other sections.

c) Output: design of maps and other graphic forms of output, also the measurement of the effectiveness of these designs.

d) Production: in accordance with the terms of reference given above, the techniques developed by the Unit are applied to practical problems, particularly but not exclusively for the constituent bodies of N.E.R.C., and the Ordnance Survey.

3. Hardware

3.1 Digitisation

Much data for which a cartographic representation is required arise as a map compiled in the field (or shortly afterwards) by a specialist in the particular topic of interest. A primary requirement is therefore to convert this analogue map into a digital form suitable for processing on a digital computer. To accomplish this task the Unit is equipped with a D-mac pencil follower. On this device a draughtsman traces a stylus manually over points or lines of interest on the map. The stylus contains a small coil emitting an electrical signal. Beneath the surface of the map table is a small aerial system mounted on a gantry. The aerial is positioned by a servo control system so that it always follows directly beneath the draughtsman's stylus. The position of the stylus is converted to digital co-ordinates by shaft encoders driven by fine wires attached to the gantry. The digital co-ordinates are recorded on magnetic tape. In addition to co-ordinate data, it is also possible to record identifying information, in the form of digits and letters, entered on an electric type-writer.

The technical specification of the table is :-

Size	-	1 metre square active area.
Resolution	-	0.1 millimetres on both axes.
Operation	-	both individual point and continuous line mode. In line mode the stylus position is sampled at a constant frequency as determined by a manually variable clock rate.
Output	-	Incremental tape recorder; 7 track, 200 characters/inch; BCD characters in even parity.

The Unit is about to exchange this table for two of a more recent design with an improved resolution of 0.05 millimetres.

3.2 Computer

The Unit is equipped with its own PDP9 computer, manufactured by the Digital Equipment Corporation. This is a small process control computer, which was acquired to control the draughting table. However, in practice it has proved convenient to use it also for much program development work.

The Unit's configuration is :-

Core store : 16 K words (18 bits, 1 μsec cycle time)
Hardware arithmetic unit
Slow tape units: 3 DEC tape units
Industry-compatible tape units: 2 units, 7 track, 200/556/800 character/inch recording
Disc: 262 K words (18 bits), 17 μsec average access time
Display: Tektronix 611 storage display.

(The Unit has on order a PDP15 (the program-compatible successor to the PDP9) with 16 K core store, 3 DEC tapes and an additional industry-compatible unit.)

3.3 Draughting table

The basic draughting table is produced by Aristo of Hamburg, whilst the electronic controls are supplied by AEG of Frankfurt. The table may only be driven on-line, direct from the PDP9 computer. There are no facilities for driving the table directly from magnetic tape or other such peripheral.

The table is 1.5 metres square. The repeatability of positioning is ± 0.01 millimetres, and the maximum absolute error, defined by taking the difference between actual and expected distances between two points is ± 0.05 millimetres. The maximum draughting speed on long straight lines is 4 metres/minute. But on the curved lines which typically appear on maps, speed has to be reduced to 53 centimetres/minute if full accuracy is to be maintained.

The table provides for 3 basic forms of reproduction:-

a) Rapidograph pen. Changes between pens of different colours must be made manually. Pens are generally only used for "quick-look" drawings.

b) Scribing. A sapphire or steel point may be used to scribe a clear line on opaque plastic sheet. This gives a fine line of high quality and is commonly used to produce (via a copy onto "peel-coat" material) open-window masks for large areas of colour. The only line thickness available is 0.1 millimetres. (When set up to draw by pen or scriber, the equipment is sometimes referred to as the "Geagraph".)

c) Light-spot projector. It is also possible to draw by projecting a beam of light onto photographic film. The beam of light is shaped by transmitting it through a mask to form the shape of a symbol. 50 symbol masks are positioned round the periphery of a disc, and any one of these symbols may be selected by rotating the disc under computer program control. (Symbol discs may be changed manually. A symbol position may be left blank, and a new symbol, reproduced on film, stuck in position over the blank.) A symbol may be projected either as a point symbol in a fixed position, or as a continuous line by moving the projector whilst light is transmitted through the symbol. A rectangular slit symbol is generally used for drawing a line. Line thickness can not be changed over a continuous range, but may be varied by selecting different symbols. The light intensity used to draw lines is varied automatically by the hardware to maintain constant film exposure at varying drawing speeds. No special action is required on the part of the controlling computer program. Double lines may be drawn by using a symbol consisting of two rectangular slits. The lines are automatically "steered" by the hardware so that they are drawn with constant separation. This is achieved by projecting the beam of light through a dove-prism which is rotated in proportion to the change in direction of drawing. It is possible to draw pecked lines under direct hardware control. Nine pre-wired pecking patterns are available for selection by the controlling computer program. (Alternatively it would be possible to produce any variation of pecking by providing a special control program to gate the light source on and off at the correct points along the line. However, it would then probably be necessary to reduce the draughting speed to allow the control program to keep pace.) When set up to draw with the light spot projector, the equipment is sometimes referred to as the "Gealux".

4. Storage of Cartographic Data on Magnetic Tape

Within the ECU system cartographic data is represented as strings of numbers, and these numbers are recorded on magnetic tape. **This** situation has two aspects:-

a) What data is recorded.

b) How the numbers are recorded on the tape

To a user working entirely within the ECU system, only (a) is important. Such a user will be satisfied that numbers are recorded on the tape, and leave the software and hardware to deal with the problems of accessing them. However, as soon as any attempt is made to transfer a magnetic tape to or from another computer system, a detailed knowledge of (b) becomes vital.

4.1 Cartographic data within the ECU system

A set of cartographic data (i.e. a "map") is stored as a consecutive string of numbers. Any number of maps may be stored on a tape, provided only that there is room for them. The term "file" (or sometimes "sheet") is used to denote the representation of the map on tape, to distinguish it from the original two-dimensional document. A file is identified by specifying the tape number and the relative number of the file within that tape - e.g. Tape 104, file 2.

Each file starts with a "header" which comprises a summary of the data contained in the file. The header includes:

1 A permanent file title.

2 A project reference.

3. The map series from which the original document was taken.

4 The relative number of the original map within its series.

5 Version number of the data. This starts at 1 and is automatically incremented each time the data is processed by a computer program.

6 A title for this particular version of the data.

7 The tape number and file number of this file.

8 The name of the computer program which produced this file.

9 The name of the operator who ran the program.

10 The date on which the program was run.

11)
12)
) As items 6 to 10 but describing the previous version of
13) the data from which the current file was produced.
14)
15)

16 Nominal scale of the data.

17 Scale factors to be applied along the x, y axes to restore the co-ordinates to those of the original document from which the file was digitised.

18 "Filter distance" which was applied to "thin-out" line data.

19 Projection in which the co-ordinate data is expressed.

20 Where the nominal position of the pole has been changed, the geographicals of the new pole expressed in the old system.

21 Geographical co-ordinates of the 4 basic corner points.

22 Geographical co-ordinates of the local origin.

23 Draughting table co-ordinates of the 4 basic corner points.

24 X, Y off-sets of file co-ordinates relative to the local origin, in table co-ordinates.

25) Where projection has been changed, as items 21, 22 expres-
26) sed in old projection.

27 Where a file represents a mosaic of smaller sheets, the table co-ordinates of the junction lines of the mosaic.

The remainder of the file consists of co-ordinate data. Co-ordinates are expressed in draughting table co-ordinates (an alternative would have been geographical co-ordinates) to an accuracy of $\pm$ 0.02 millimetres over a maximum area of approximately 4.8 metres square. Co-ordinates are related to individual points and to lines. A line is represented as a string of points. These line points are in effect sample points separated by a small distance (generally 0.25 millimetres or less). Thus a map is stored as a string of points. The types of points provided for are:-

1) Corner points. A map may have any number of corner points, or none at all. If there are corner points, these need not be at the physical corners. However, it is most convenient if a map is rectangular with 4 corner points at the true corners.

2) Individual points. These are individual data points within the map. There may be any number.

3) Start of line point
4) Internal line point
5) End of line point

A line is represented as a start of line point, followed by any number of internal line points, followed by an end of line point.

The co-ordinates of a map are generally specified relative to the bottom left corner of the map. However, it frequently happens that a large area is first processed as a mosaic of small maps. (The mosaic is usually, but not necessarily, a regular rectangular lattice.) It must be possible to reproduce the individual maps in their correct relative position in the mosaic. To facilitate this, a "local origin" may be specified for the map series. This origin is usually the bottom left corner of the whole series. The displacement of the individual map from the local origin is stored in the map header (item 24 in the list above). Thus the map co-ordinates relative to the mosaic may be obtained simply by adding the local origin off-sets to all the co-ordinates stored in the body of the file.

Every corner point, individual point, and start of line point may be labelled with "feature codes". Each point may have up to 50 codes, each code taking the form of a 5 digit integer (positive or negative). Just what a feature code represents is determined by the user. Typical code assignments are:-

(a) height of point above sea level

(b) rateable value of heraditament situated at a point

(c) two codes associated with lines, signifying the type of soil to be found on either side of the line

(d) heights of contour lines above sea level.

4.2 Communication of data with other computer installations

Where data on magnetic tape is sent from one computer installation to another, it is important for the receiving installation to know not only what information is recorded, but exactly how it is recorded. Appendix I to section 4 is a copy of the proposals for a standard tape format which have been submitted by the Unit to the ICA.
Appendix 2 to section 4 gives details of additions which the Unit has made to the original proposals. These additions allow for the communication of "figure fields".

Section 4 - Appendix 1

A Standard Format for the Exchange of Cartographic Data on Magnetic Tape

It is commonly agreed that the digitising of cartographic data is a costly and time consuming practice. It would be a great waste of valuable resources therefore to digitise the same information more than once. If this is to be avoided there must be some straightforward means for the exchange of data between different organisations. This note attempts to define a standard format for data exchange.

The requirement to define a standard medium for the exchange of data does not mean that this data must be stored in a common format within all computing systems. The study of automated cartography has not progressed sufficiently to make it possible to recommend a standard design of data bank, indeed this may never be possible. The approach has been rather that information can be standardised just after the stage where it has been digitised, checked, and if necessary corrected, but before the point where it is assimilated into a data bank.

For this reason, it is anticipated that co-ordinate data exchanged by these means will usually be specified as the table co-ordinates taken direct from the digitising process. It will be up to the organisation receiving the data to apply the necessary transformation into the units which it requires for assimilation into its own data bank.

The transmission of data in the form directly as taken from the digitiser has the advantage that it should be possible to recover the original map exactly (to the accuracy of the digitisation process). However there is no reason why the organisation issuing the data should not apply any arbitrary transformation which it feels is useful.

Examples of such transformations are:-

1. Rotation of the digitiser co-ordinate system so that its axes are parallel to the natural axes of the map.
2. Transformation of all co-ordinates into latitude and longitude (with a suitable change of scale and origin to enable the resultant numbers still to be expressed in the standard format).

1. PHYSICAL TAPE SPECIFICATION

1.1 7-track 'IBM-compatible' magnetic tape

Punched cards and paper tape would be too bulky.

Most computer installations, even non-IBM installations, have some provision to use 'IBM-compatible' tape. Newer installations are generally biased towards 9-track tape, but even these make some provision for handling 7-track tape. This document therefore defines a standard for 7-track 'IBM-compatible' magnetic tape. (However, it is anticipated that this situation will need to be reviewed if other media come into international use.)

1.2 Binary-coded-decimal character recording

Straight binary recording would make for more efficient use of tape. However, the use of character recording has the advantages:-

(a) Data may be accessed directly by standard FORTRAN programs. (Basic machine code programs may be developed where local circumstances make this necessary.) The FORTRAN character code is given in Appendix I. This is the original IBM FORTRAN 6-bit code.

(b) In the event of uncertainty it is always possible to list the tape on a line printer, and decipher the results visually.

It is felt that the use of BCD characters, plus the normal hardware parity checking make it unnecessary to add further error protection (e.g. insertion of checksums).

1.3 Even parity

This is normal practice where BCD recording is used.

1.4 Recording density 556 characters/inch

Alternatives are 200 or 800 characters/inch. The higher the recording density the more efficient the use of tape, but the more susceptible the recording process becomes to hardware errors. 556 characters per inch is the best compromise. It is recommended that installations with a choice of recording densities should make provision to use the alternatives if necessary.

1.5 Record length a multiple of 120 characters

For simplicity it is suggested that data should be recorded in units of information (sometimes termed "logical records") of 120 characters. It is important for diagnostic purposes to be able to list the data on a line printer. A common carriage width for printers is 120 characters, though some printers allow up to 160 characters.

It is recommended that in general each physical record on the tape should comprise a single logical unit of 120 characters. However, since the standard inter-record gap for 7-track tape is 0.75 inches, such a short record length may lead to inefficient use of the tape. Where large quantities of data are involved, a number of units of information may be combined to form each physical record, provided that every record has exactly the same length. (The number of units of information per physical record should be specified in a covering letter accompanying the magnetic tape itself, as well as in the header information.)

2. FORMAT SPECIFICATION FOR UNITS OF INFORMATION

Three different formats for the 120 character units of information have been defined:-

2.1 Header information

The first 10 units of information describing a map comprise information specifying general features of the cartographic data which follows on the tape. This information can be in free format (i.e. accessed by FORTRAN via the 'A' or character format), and hence may only be interpreted by printing it and deciphering it visually.

It has not been possible to define the format any more precisely, since the requirements are likely to vary from case to case, but it is clear that the information supplied should be as comprehensive as possible. Such information as the source document from which digitising was performed, the date and method of digitising, area covered, scale, etc. will be required. The geodetic datum on which horizontal and vertical co-ordinates are based should normally be specified in adequate detail. In view of the possibility of future re-adjustments this specification should normally be supported by a date, and the name of a particular co-ordinate list, or the actual co-ordinates of some specified points. Sources for data on population, wheat yield, rainfall, etc. should be given in similar detail.

2.2 Single point information

These units of information are related to point, as opposed to continuous line, information.

Character	Information
0 - 1	Mark number - denotes the type of information described by this unit. Mark numbers so far assigned are: 03 Start of line. A start of line unit <u>must</u> be followed by units describing the course of the line (section 2.3) 04 Individual point 05 Cornerpoint. Practical experience has shown it important that a map should contain a number of reference points, especially where several maps are later to be juxta-positioned. Usually these will be the 4 corners of a bounding rectangle. 06 End of map. The following unit is the start of information for another map. 30 End of tape. As for end of map, but no other maps follow.
2 - 3	Correction code. This is used where the current data is intended to update earlier information. This is discussed in section 2.4 below.
4 - 9	X co-ordinate. The range of co-ordinates is -99999 to 999999. No attempt is made to specify the units in which the co-ordinates should be given, as these may vary from one application to another. The unit of measurement should be specified in the header information.
10 - 15	Y co-ordinate.
16 - 17	Number of feature codes which follow immediately.
18 - 23 24 - 29	Maximum of 17 6-digit feature codes. The significance of each code must vary from application to application. The use of these codes should be defined in the header information. e.g. 1st code = height above specified sea level datum in metres. 2nd code = annual rainfall for 1962 in millimetres. In particular it is permissible to use one or more pairs of feature codes to extend the range of accuracy

of the x, y co-ordinates. For example codes 1 and 2 (digits 18 to 23 and 24 to 29) could be used to expand x, y to 12 digit accuracy. A further feature code might be used to give an indication of the expected re-
114-119 liability of the data.

2.3 Line information

A start of line unit information must be followed by further units defining the course of the line. To make efficient use of the tape, the line will be specified as a string of relative co-ordinates, denoting the change Δx, Δy in x, y from the previous point. Each 120 character unit will be made up of 30 pairs of Δx, Δy relative co-ordinates, each of 2 digits. The range of each relative co-ordinate value will be 01 to 99, with a shift of origin, so that 50 represents the zero increment of position.

e.g. number on tape	relative co-ordinate increment
52	+2
51	+1
50	0
49	-1
48	-2
47	-3

Zero values for shifted Δx, Δy (i.e. 00 00 on the tape) will be reserved to indicate the end of line. The remainder of the record will also be packed out with zero values.

It is recognised that this format may not produce an economical method of data storage in the special case where there occurs a large proportion of straight lines. Whether or not an additional format for straight line information is desirable will become clearer with experience.

2.4 Correction codes

The correction codes are intended to facilitate the provision of a data tape which may be used to correct or update a tape which was produced at an earlier date. (Correction codes are not intended to supply corrections to data supplied elsewhere on the same tape. All data tapes should be checked for internal consistency and accuracy before being transmitted to another organisation.) If a tape is a correction tape then all data is assumed to be correction data, and every correction code should be non-zero. For original non-correction tapes all the correction codes must be zero (or blank). (Thus for many purposes the correction code facility may be ignored.)

The correction codes permit the giving of correction commands related to specific co-ordinate positions. E.g. 'delete the line which passes through this point.' This pre-supposes that the correction co-ordinates have been transformed to the same origin and direction of axes as the original data. It would be extremely difficult to supply correction co-ordinates which exactly matched those of the data to be corrected, and some degree of tolerance should be allowed for. This tolerance should be specified in the header information.

The correction codes so far assigned are:-

00	Not a correction (zero characters or blanks)
01	Delete line which passes through this point (within tolerance)
02	Delete point which has these co-ordinates
03	Insert this line (must be associated with mark number 3)
04	Insert this point
07	Change line feature codes to the codes given in this record
08	Change point feature codes to the codes given in this record

3. RECOMMENDATIONS

1. The above format should be accepted as the standard for the exchange of digitised cartographic information on 7-track magnetic tape.

2. There should be regular consultations between interested organisations for the extension of the standard in the light of future developments. It would be advantageous if these were held in advance of normal ICA meetings.

Appendix I

TAPE CHARACTER CODE

Number on Tape	Number in Store	Character	Number on Tape	Number in Store	Character
00			40	40	-
01	01	1	41	41	J
02	02	2	42	42	K
03	03	3	43	43	L
04	04	4	44	44	M
05	05	5	45	45	N
06	06	6	46	46	O
07	07	7	47	47	P
10	10	8	50	50	Q
11	11	9	51	51	R
12	00	0	52		
13	13	=	53		
14	14	·	54	54	X
15			55		
16			56		
17			57		
20	60	space	60	20	+
21	61	/	61	21	A
22	62	S	62	22	B
23	63	T	63	23	C
24	64	U	64	24	D
25	65	V	65	25	E
26	66	W	66	26	F
27	67	X	67	27	G
30	70	Y	70	30	H
31	71	Z	71	31	I
32			72		
33	73	,	73	33	.
34			74		
35			75		
36			76		
37			77		

All numbers are in actual representation. In most systems the exchange between representation in core and on tape is performed by hardware. This character set is a subset of the original IBM 6-bit FORTRAN character set. The set has been left as large as possible, whilst omitting those characters known to cause difficulties on input in specific installations. Tape codes for which both the number in store and printing character have been left blank should not appear on the tape.

Section 4 - Appendix 2

Standard Format for Cartographic Data - Inclusion of Figure Fields

This note suggests an extension to the original proposal for "A Standard Format for the Exchange of Cartographic Data on Magnetic Tape". The earlier proposal was concerned with data ("feature codes") associated with individual unique physical positions specified by x, y co-ordinates. Where a number of different positions were involved, it was envisaged that the co-ordinates for each point would be specified. However, on some occasions it is convenient to transmit data for a large group of related points, where each point lies on the intersection of a regular and usually rectangular grid or matrix. The corresponding matrix of data values or "feature codes" is sometimes termed a "figure field". In these circumstances it is wasteful to transmit the x, y co-ordinates of every point, since the position of the points can easily be inferred from the order in which the points are recorded, provided only that the position of one corner of the grid and the row and column spacings have also been specified.

A figure field will be specified as a single logical unit of 120 characters specifying the corner co-ordinates of the grid and the row and column spacings, followed by as many units as may be necessary to specify the data values. The first unit will have the standard format for single point information (section 2.2 of the earlier note), employed as follows:-

Character	Information
0 - 1	The mark number 07 will indicate that figure field information follows in the succeeding logical records. This mark number may be written in characters 0 - 1 of the first record following either 1) the last record of the set of 10 header records which begin a map sheet, 2) a line information record which contains the end-of-line co-ordinate pair 00 00, 3) an individual point (mark number = 04), or 4) a corner point (mark number = 05).
2 - 3	Correction code (cf. section 2, 4) The following correction codes will be used in connection with figure fields, but not necessarily in records with mark number = 07.

	00	Not a correction. Used on data tapes.
	05	Delete figure field having this lower left hand corner. No figure field will follow this record and mark number 04 should be used.
	06	Insert this figure field. The mark number should be 07 and figure field data should follow.
	09	Change the feature codes in the mark number record of the figure field with this lower left hand corner to those given here. Use with mark number 04.
	10	Change the lower left hand corner of the figure field with this lower left hand corner to the co-ordinates given here in feature codes 1 and 2. Use with mark number 04.

4 - 9	X co-ordinate of the lower left hand (SW) corner of the figure field.
10 - 15	Y co-ordinate of the lower left hand corner of the figure field.
16 - 17	Number of feature codes which follow in <u>this</u> record. The first eleven feature codes have unalterable meanings and must be given. Hence this number should never be less than 11.
18 - 23	Integer part of the width (E-W distance) of the figure field.
24 - 29	Fractional part of the width of the figure field.
30 - 35	Integer part of the height (N-S distance) of the figure field.
36 - 41	Fractional part of the height of the figure field.
42 - 47	Number of columns N-S vectors) in the figure field. This number is one greater than the number of intervals into which the width (E-W distance) is divided.
48 - 53	Number of rows (E-W vectors) in the figure field. This number is one greater than the number of intervals in which the height (N-S distance) is divided.

54 - 59	Number of codes attached to each grid node. The format for these codes is given below.
60 - 65	Integer part of the difference in azimuth (radians) between the X-axis of the co-ordinate system and the bottom (S) edge of the figure field.
66 - 71	Fractional part of the difference in azimuth (radians) between the X-axis of the co-ordinate system and the bottom (S) edge of the figure field.
72 - 77	Integer part of the difference in azimuth (radians) between the Y-axis of the co-ordinate system and the left (W) edge of the figure field.
78 - 83	Fractional part of the difference in azimuth (radians) between the Y-axis of the co-ordinate system and the left (W) edge of the figure field.
84 - 89) 90 - 95) 96 - 101) 101 - 107) 108 - 114) 114 - 119)	Feature codes to be applied to the entire figure field in a fashion similar to line feature codes.

Each node point in the figure field has a fixed number of data values or feature codes associated with it. This number is specified in characters 54 - 59 of the logical unit as specified above. These feature codes are specified as 6 digit numbers, packed 20 to each of the following logical units. These units are formatted according to the following rules:

1. All codes for one node are given before any codes for the next node.

2. The first node to be listed is the lower left hand (i.e. S.W.) corner point.

3. The nodes are ordered in increasing Y co-ordinate, along X scan lines.

4. When all the points on an X scan-line are exhausted, the next point in the sequence will be the first (minimum Y) point on the next X scan line. The codes for the first point on a scan line must begin the next logical unit. Consequently some codes in the final unit of each X scan-line may be unused. No meaning should be attached to any digits of unused codes.

No limit is set to the number of feature codes associated with each data point, but it is suggested that a maximum of 20 (i.e. to fill one logical unit) be observed in practice.

The number of packed logical units following the first unit with mark number 07 may be computed from the information specified in that unit (i.e. from the number of columns, number of rows and number of codes per data point).

Six digit codes are used so that the format of the figure field data values will be consistent with the feature code format for individual line and point data. The six digit codes should be viewed simply as codes, not necessarily as numbers. It is not intended to specify rigidly how real (i.e. non-integer) numbers (such as heights) should be transmitted in all cases. However, it will frequently be convenient to pass real numbers as two codes, the first code being the integer part of the real number, and the second code the fractional part. It is recommended that this practice should be adhered to wherever possible.

5. Brief Guide to Computer Programs and Their Interrelationships

This section supplies a thumbnail sketch of the computer programs currently available, and shows how they are related. Detailed program operating instructions follow in section 6.

Programs have been given names of up to 6 characters. By convention these names are written entirely in capital letters. Sometimes these names are meaningful, sometimes their origins have been lost in the mists of antiquity. In 1969 there started a major re-write of most of the Unit's programs. Thus some of the current programs have an "N" as the initial letter of their names to distinguish the new from the old versions.

The program operating instructions include the names of the program authors. These are the authors of the current versions. Very often they were able to lean substantially on the earlier efforts of other programmers. Foremost amongst these pioneers were Herr W. Jud of AEG, Berlin, and Mr T W Thacker of the ECU.

5.1 Input programs

Figure 1 shows the programs normally used to process data as it enters the ECU system.

NMTTDT processes a data tape produced on the D-mac digitiser, converting the data to the standard internal ECU format described in section 4.1. Feature codes may be entered on the D-mac typewriter in a convenient shorthand form, and these are expanded by the program.

NJOIN reads the tape prepared by NMTTDT and prepares a corresponding output tape in which the map co-ordinates are rotated and translated so that the South-West corner is at the origin and the South border lies along the X axis. The program asks the user to specify the nominal co-ordinates (e.g. the latitude and longitude) of the map corner points. By relating the nominal co-ordinates to the actual co-ordinates it is possible to make some allowance for any distortion of the original document. The program also computes the off-set of the map specifying its relative position within a mosaic of maps. Nine different map projections are catered for.

NSJOIN provides the same facilities as NJOIN for maps on a regular rectangular grid projection (e.g. British National Grid) or on non-standard projections.

Data may be converted to digital form at some other installation. The programs which accept a magnetic tape formatted according to the standard for data exchange and convert it to internal ECU format are described in section 5.4. The converted tape may be accepted directly or in some cases further processed by one of the JOIN programs to provide the requisite co-ordinate conversions.

5.2 Output programs

NGEAG is the basic program for the reproduction of maps on the Geagraph (i.e. the high accuracy plotter with pen or scriber). The map may be reproduced at any scale, and any specific items omitted or included by nomination of the corresponding feature codes. The feature codes may also be used to draw different items with different coloured pens. The map can be drawn either at the table origin, or at its correct relative position within a mosaic of maps. It is possible to reproduce either a whole map (or mosaic of maps) or merely parts of a map (or maps) falling within a specified rectangular window.

Where a map is played back at a greatly reduced scale some generalisation of the linework can be provided. Map A illustrates this in a map produced for the National Institute of Oceanography by scribing with a 0.1 mm. unheated scriber. The original material was digitised at a scale of 1:1 million. It is reproduced here with varying degrees of generalisation, at a scale of 1:20 Million.

NGEAL provides the equivalent facilities of NGEAG on the Gealux (i.e. the high accuracy plotter with light-spot projector). It is possible to select both point and line feature symbols according to feature code. With line symbols a line which is either continuous or pecked in one of the 9 pecking patterns provided by the hardware can be selected, and with point symbols the symbol can be projected at any selected orientation. Map B illustrates the output of NGEAL and is a part of a map of the Bideford area, produced in conjunction with the Ordnance Survey. Map C shows additional output, of geological boundaries in the Abingdon area. These were produced for the Institute of Geological Sciences.

NSD provides the equivalent facilities of NGEAG on the computer storage display. This is useful for a "quick look", as drawing times are very much shorter than on the high accuracy plotter. The map reproduced in Map D was drawn on the display and then photographed, as was the oblique view of the same area.

NGEANA draws "3 dimensional" maps suitable for viewing through red-and-green spectacles, or as stereo-pairs. Variations of this program are available to drive both the Geagraph and the Gealux.

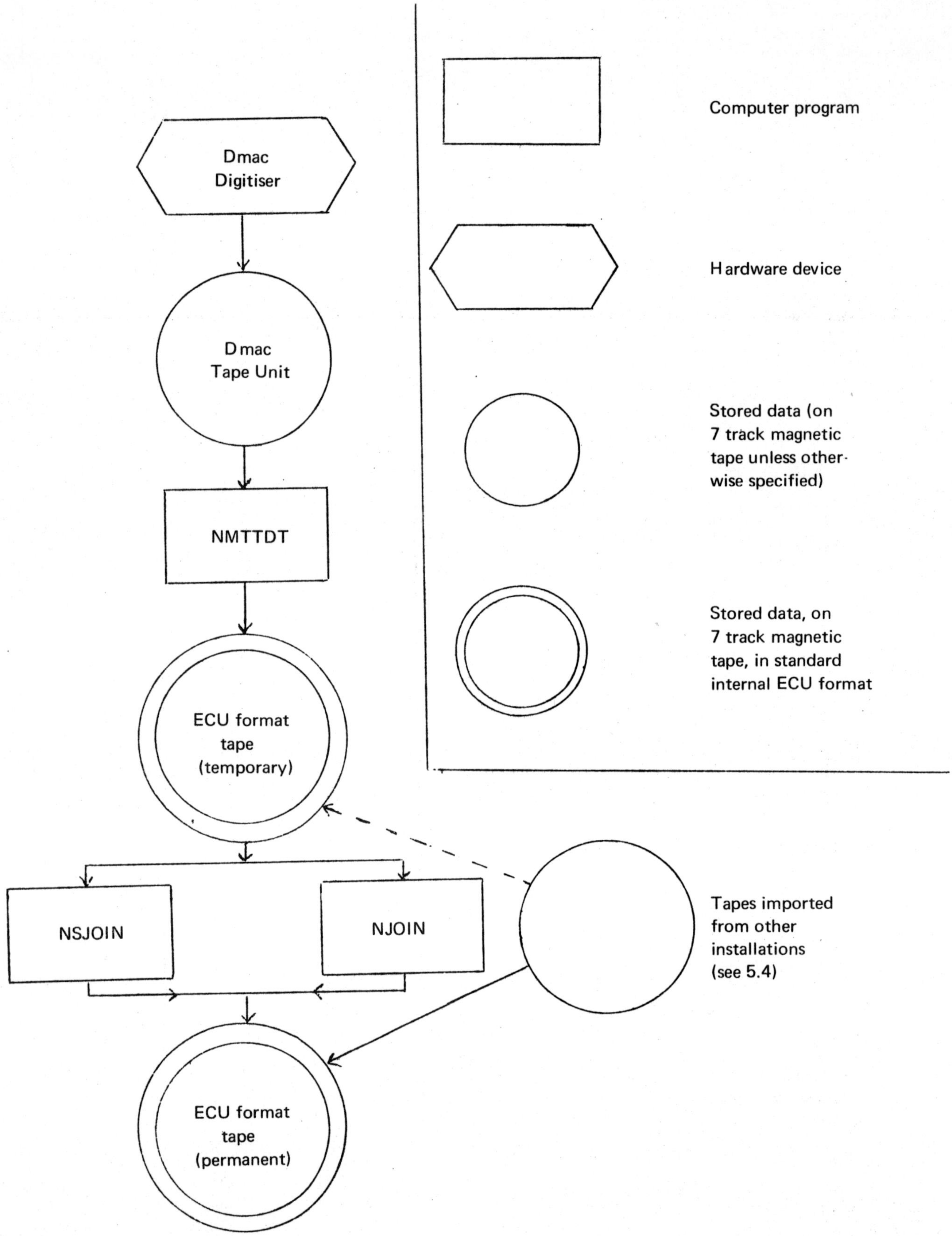

Figure 1 — Data input

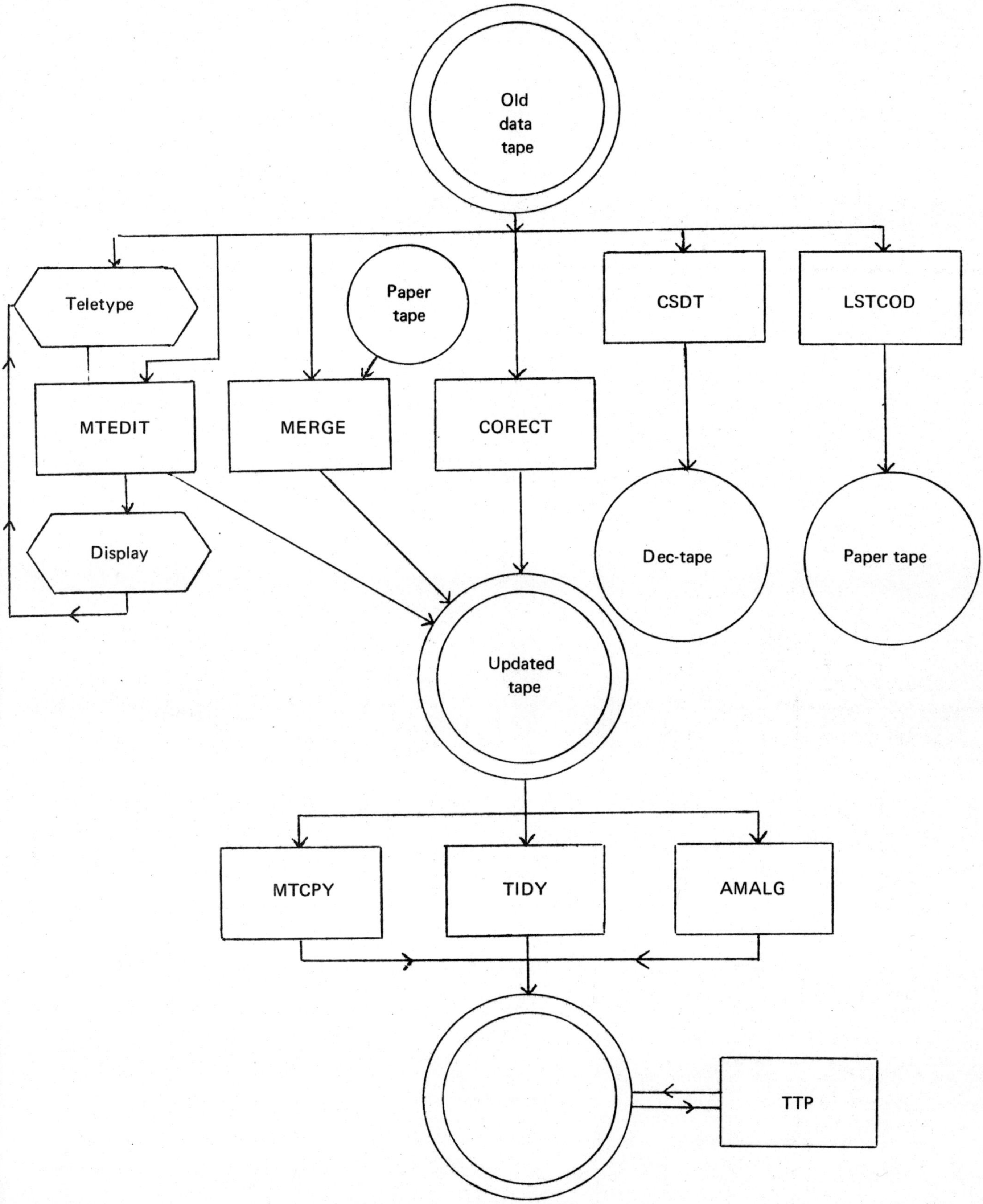

Figure 2 – Correction and Utility Programs

LINLAB automatically labels lines with their corresponding feature code values (e.g. the heights of contours). The labels are produced by using the digit symbols on the Gealux.

LPP and SPP are provided to assist in the design of line and point symbols on the Gealux. Symbols can be produced in various combinations, under control of data read from paper-tape, to build up a symbol catalogue.

5.3 Data correction and utility programs

The flow of data between these programs is illustrated in Figure 2.

CORECT permits the correction of both positional and feature code data as stored on magnetic tape. A tape of corrections must first be prepared on the D-mac digitiser. The user is able to enter such directives as "delete the line which passes through the point which has just been digitised" or "change the feature codes of the point at the position just digitised". The correction tape should be processed via NMTTDT and JOIN in the normal way. CORECT then merges the old data tape and the correction tape to produce an updated tape. (Because only two 7-track tape units are available, the correction data must be copied to one of the small DEC tape units using program CSDT, before CORECT can be run.)

MTEDIT allows selective correction on an interactive basis. The old map data is reproduced (on the display or the Geagraph) one line or point at a time. The recorded feature codes of the item are output on the teletype. The user is given the option to accept the item, to accept the item but change the feature codes, or to delete the item.

MERGE is particularly applicable to point data, (though it can also be employed with line data), and is used to ease the burden of entering feature codes. The points are first digitised without regard to feature codes. The feature codes are then punched up off-line on paper tape, and merged with their corresponding points by means of MERGE. (It is necessary to know the order in which the points have been digitised. This is achieved by reproducing the points on the Geagraph, via NGEAG, with the pen held in the permanently down position, thus recording the movement of the plotter between the successive points.)

LSTCOD may be used to aid in the off-line checking of feature codes. For each item on the map its co-ordinates (i.e. single co-ordinates of a point, start and end co-ordinates of a line) and its feature codes are punched on the high speed paper tape punch for subsequent printing and checking.

MTCPY is a utility program which may be used to copy and verify magnetic tapes. This program will work not only for tapes recorded in the ECU internal format but for tapes recorded in any parity, density and record length.

TIDY will re-order tape files. This is useful where many changes and corrections have taken place, to discard old data and restore the remaining files to their logical order in a map-series.

AMALG amalgamates the separate files of a mosaic of maps into one large file containing the whole series. This is particularly advantageous where the maps are to be reproduced at a reduced scale, as it is possible to filter the data points to the lower density which will be needed for reproduction.

TTP is used to recover from any error (of hardware, software or operator) which causes the computer to stop processing mid-way through the production of a tape file. A correct end-of-tape marker is placed at the end of the last complete file.

5.4 Processing of tapes in standard exportable format

The flow of data between these programs is illustrated in Figure 3.

STG converts data from internal ECU format to exportable format but still employing the checks and markers peculiar to the DEC software system.

FFOUT performs a similar function to STG but converts a tape containing a single figure field to exportable format within the DEC system. STNDFF performs the same task for a tape containing multiple files of figure fields.

IBMOUT will remove the checks and markers peculiar to the DEC system, producing a "clean" tape suitable for export.

Data for import follows the reverse path. IBMIN adds the checks and markers to the "clean" tape to make it compatible with the DEC system.

Programs STNDRD and FFIN then convert the intermediate tapes to standard internal map files and binary figure fields respectively. (It is currently assumed that an exportable tape holds either normal point and line maps or figure fields but not both. This is not an essential feature and the system could be made more general if required.)

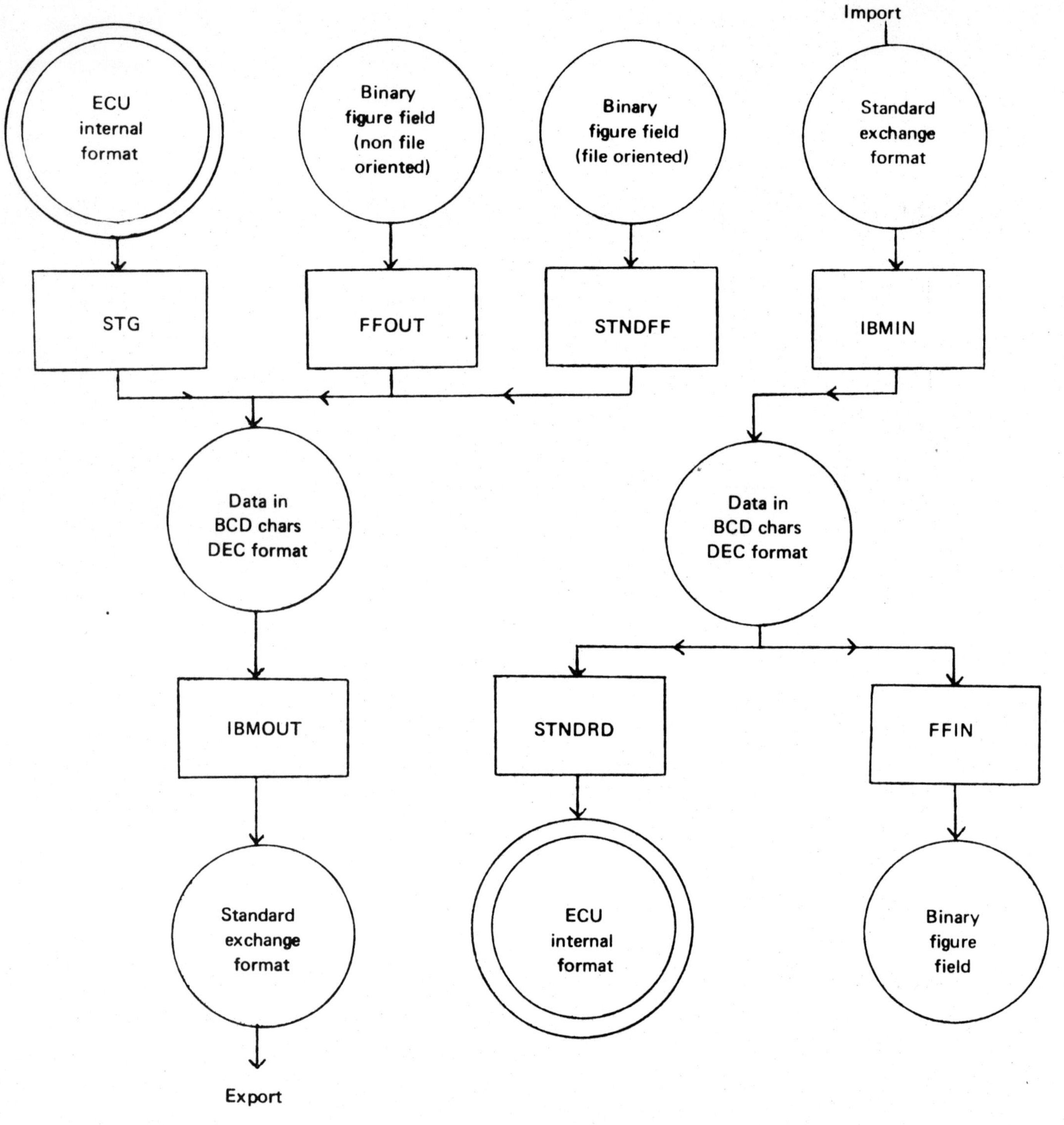

3(a) Data tape export

3 (b) Data tape import

Figure (3) Standard Data Type Processing

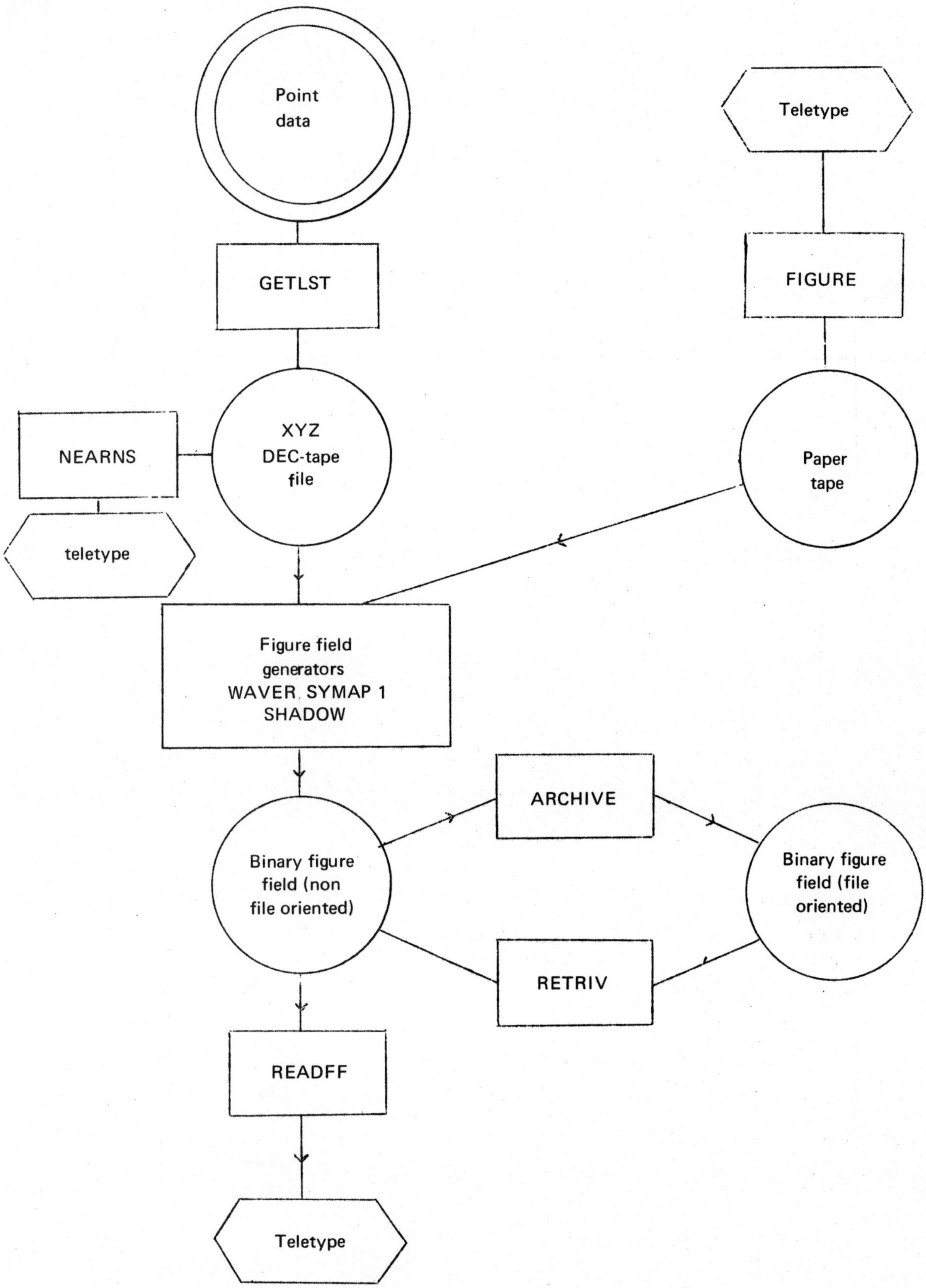

Figure 4(a) — Figure Field Generation

5.5 Figure fields and contouring

Much data is first collected in the form of sample values at irregularly or randomly located points. Examples so far processed by the Unit including soundings taken during oceanographic survey, the height of chalk strata measured at well sites, and chemical concentrations as measured in stream sediment. The sample values are frequently termed "Z" values, implying a measurement in the 3rd dimension, in addition to the 2-dimensional co-ordinates of the point location. For both cartographic and computational purposes it is frequently more convenient to deal not with a collection of irregularly distributed points but with a regular rectangular grid of points. A rectangular grid of points with its associated "Z" values is termed a "figure field". Thus a general requirement is a program to process the original irregularly distributed data points and interpolate likely "Z" values at the regular points of the figure field. The data flow for this process is illustrated in figure 4 (a).

GETLST takes data stored in the internal ECU format and converts it to a string of X, Y, Z values stored in character form on DEC tape. This is more convenient for processing by the figure fields generation programs.

FIGURE is a small program which interrogates the user and prepares a control paper tape which is used to guide the figure field generation programs.

SLOPES, WAVER, and SHADOW are three figure field generation programs. These employ different interpolation schemes and require different amounts of computer time. The user should select, with the guidance of members of the Unit, the program most suitable for his particular application. SLOPES employs an interpolation scheme very similar to that developed by Shepard for the Harvard SYMAP package. This constrains the generated surface with computed data point gradients, but can accept only 1350 data points. WAVER does not utilise data point gradients, but can accept 3000 data points. SHADOW may be used where steep gradients might otherwise be lost within a coarse figure field grid.

The operation of all 3 programs is detailed in section 6 under the title of SLOPES. It is hoped that by the time this document is published, these programs will also be available to computer users at Imperial College, on the CDC 6600 of the University of London.

The output from the figure field generation programs is a single figure field on magnetic tape. ARCHIVE may be used to file a number of figure fields on the same tape and allow reference to each by name. RETRIV will restore any named figure field file to a tape of its own. READFF prints out the "Z" values of a figure field for diagnostic purposes.

When judging the reliability of a figure field, the user may wish to obtain a measure of the randomness of the spatial distribution of the original data point set from which the figure field was generated. The program NEARNS provides this by evaluating the nearest neighbour statistic for the data points.

Once a figure field has been produced it may be further processed in a variety of ways. A frequent requirement is the production of a contour map of the surface described by the figure field. This may be accomplished by means of one of the three CTOUR programs. Three variations - CTOURI, CTOUR2 and CTOUR3 are available. They all produce the same contour map, but employ different computer peripherals, as illustrated in Figure 4(b). CTOUR1 and CTOUR3 will accept either file oriented or non-file oriented tape input, CTOUR2 accepts only non-file oriented tapes. CTOUR1 is the normal production contouring program, and is designed to use the disc as efficiently as possible. Output will normally be in standard format on magnetic tape, but DEC tape and the storage display may also be used. CTOUR2 provides back-up for CTOUR1, and is capable of working without the disc. Output direct to the Geagraph is also possible. CTOUR3 is a "quick look" program producing output only on the storage display. (Each contour elevation for a 50 row x 50 column figure field takes only 30 seconds.)

As an illustration of these facilities Map E shows a contour map for a point data set of gravity values.

5.6 General processing programs

PROJCH will convert a map stored in internal ECU format from one projection to another (9 different projections are catered for). It is also possible to change the position of the poles as used for the output projection.

GRATGN will generate a map file of graticule lines in any one of the 9 projections also processed by PROJCH. Map F shows the Gulf of Aden map also shown in Map A, but with the projection changed and graticule lines added.

SAMPLE is used to determine the lengths of line making up a map. Point densities and line lengths may be evaluated either for individual items or for items grouped by feature codes.

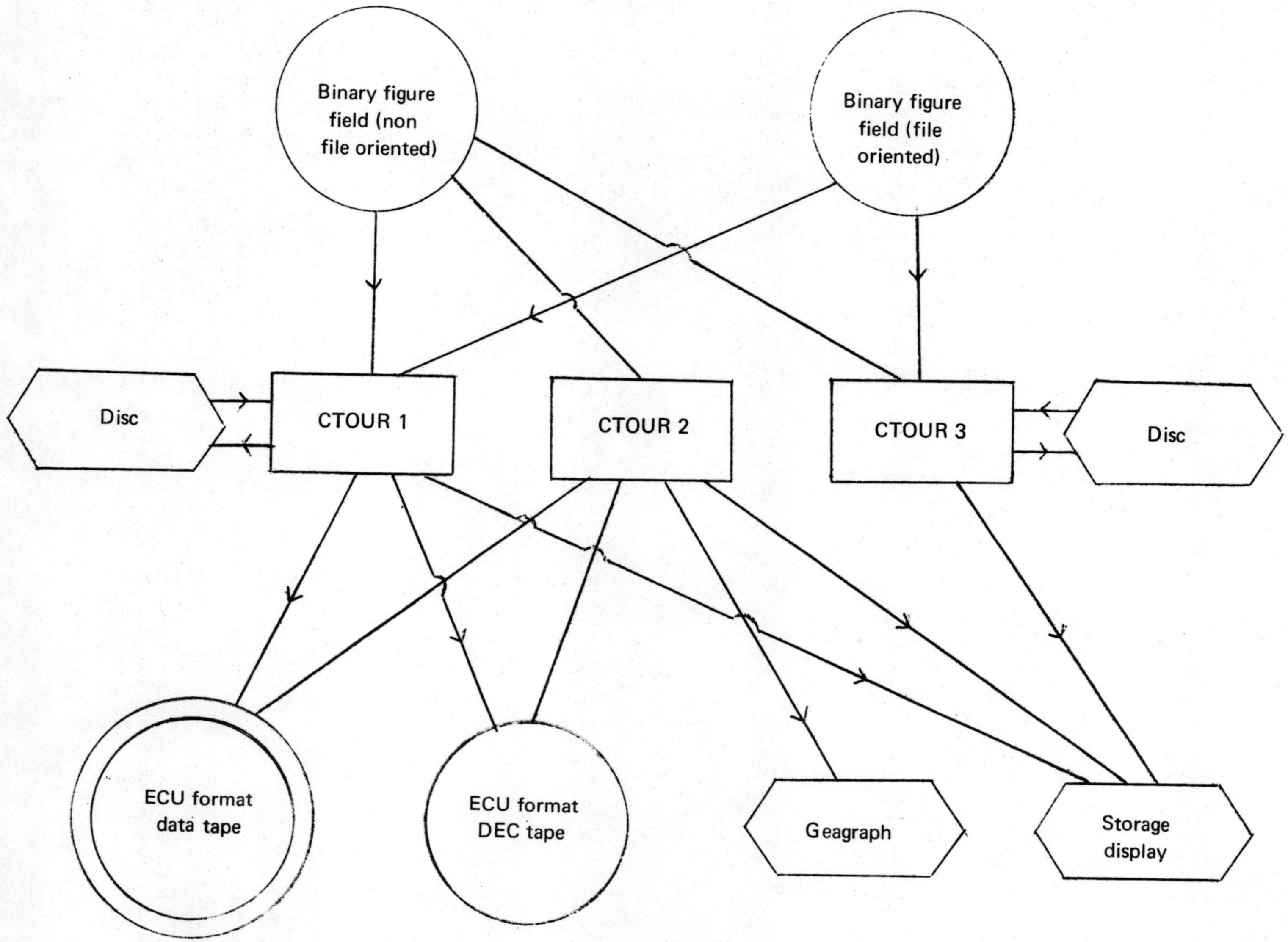

Figure 4(b) — Figure Field Contouring

Map A. Map of the Gulf of Aden
showing varying degrees of generalisation

Scale 1 : 20 million (See page 26)

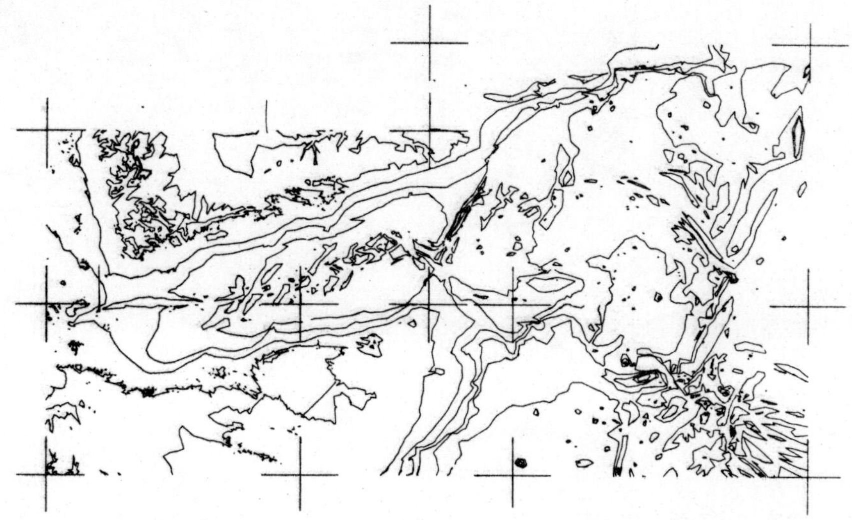

Tolerance 0.1 mm

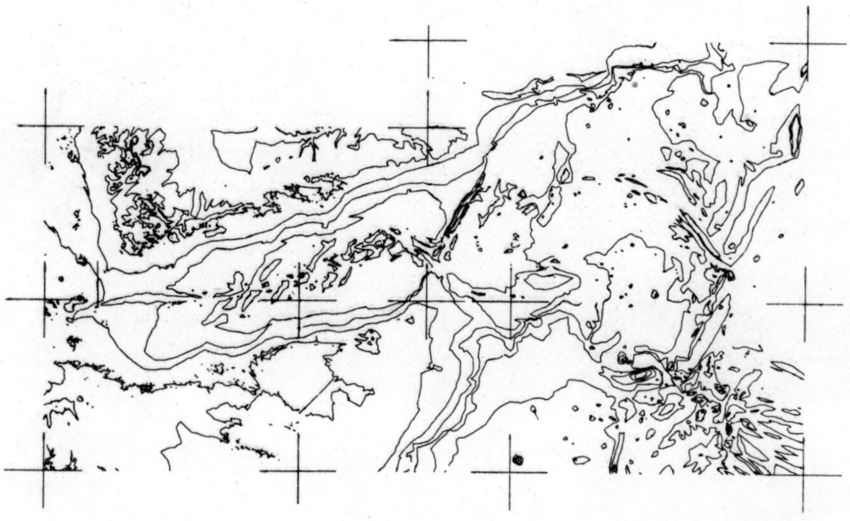

Tolerance .05 mm

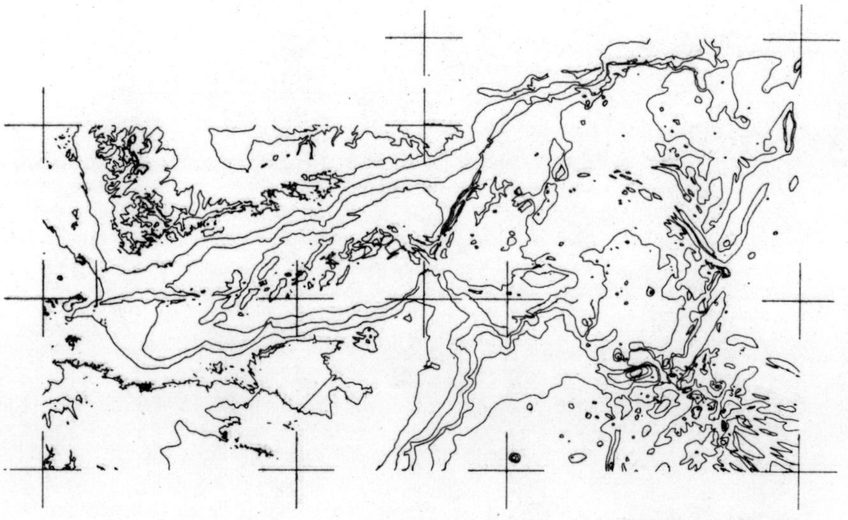

Tolerance .01 mm

Map B. Part of the Bideford area showing Light Spot Projector Output
Scale 1 : 63,360
(See page 26)

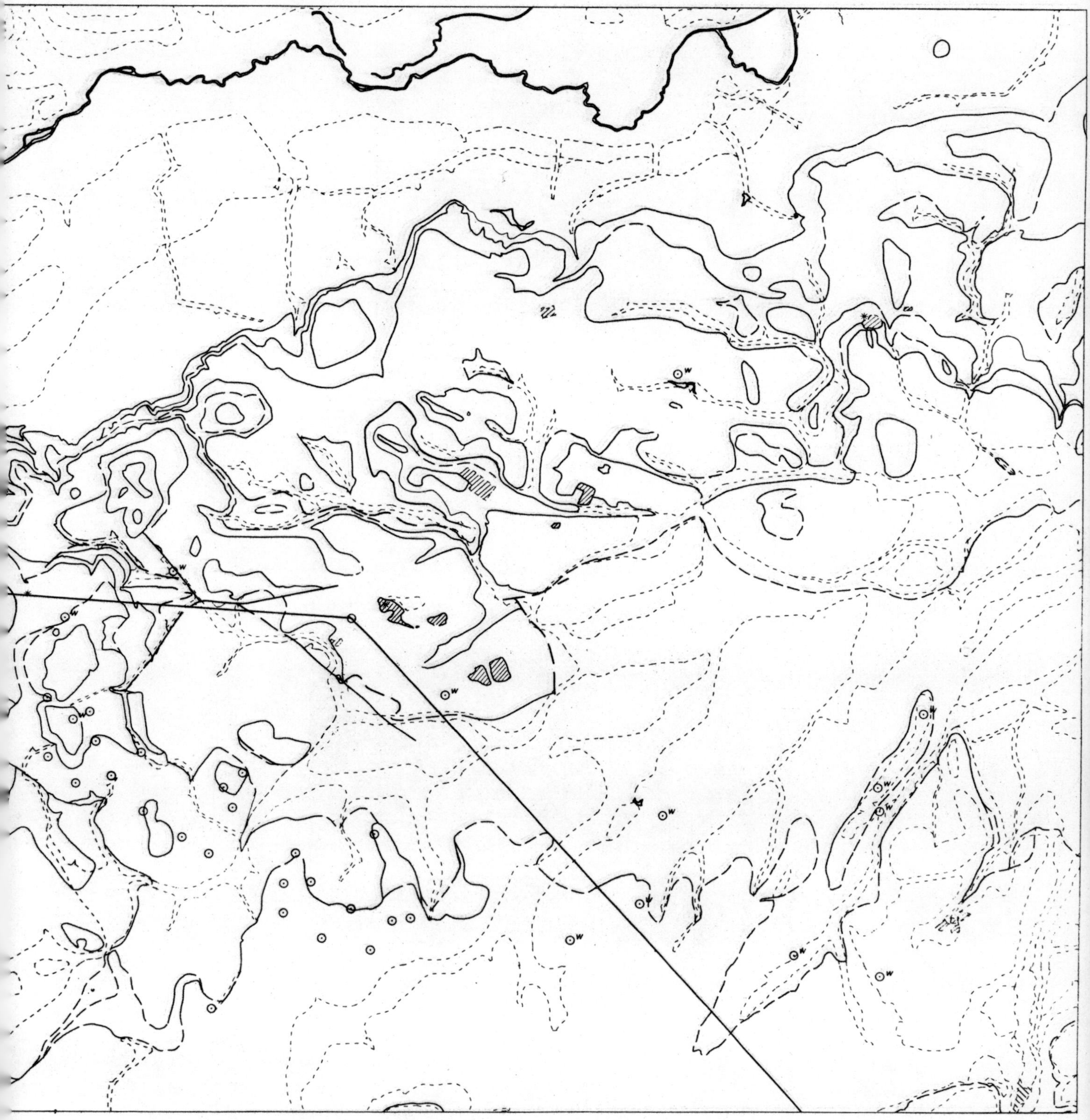

Map C. Geological boundaries in part of the Abingdon area.
Scale 1: 63,360
(See page 26)

Map D.

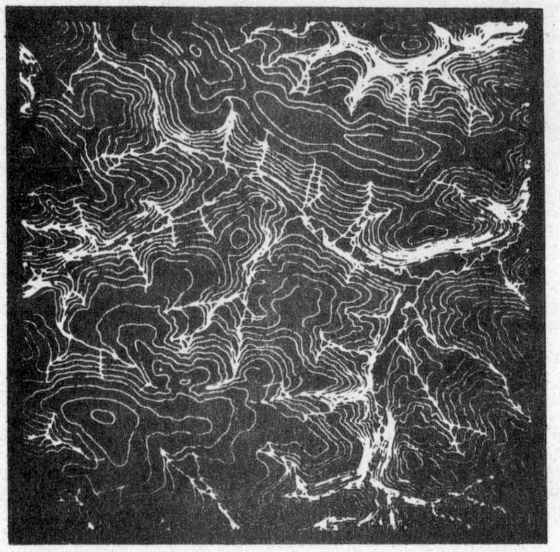

Cathode Ray Tube display of topographic contours in S.W. quadrant, Bideford area

(See page 26)

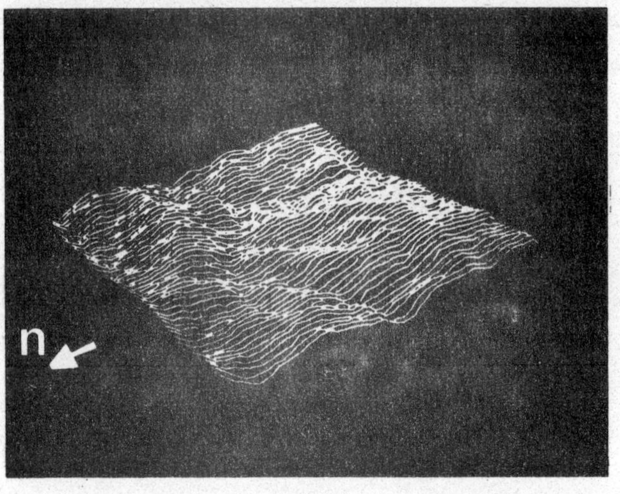

Perspective view of the S.W. quadrant of the Bideford area, viewed from the N.W.

(See page 26)

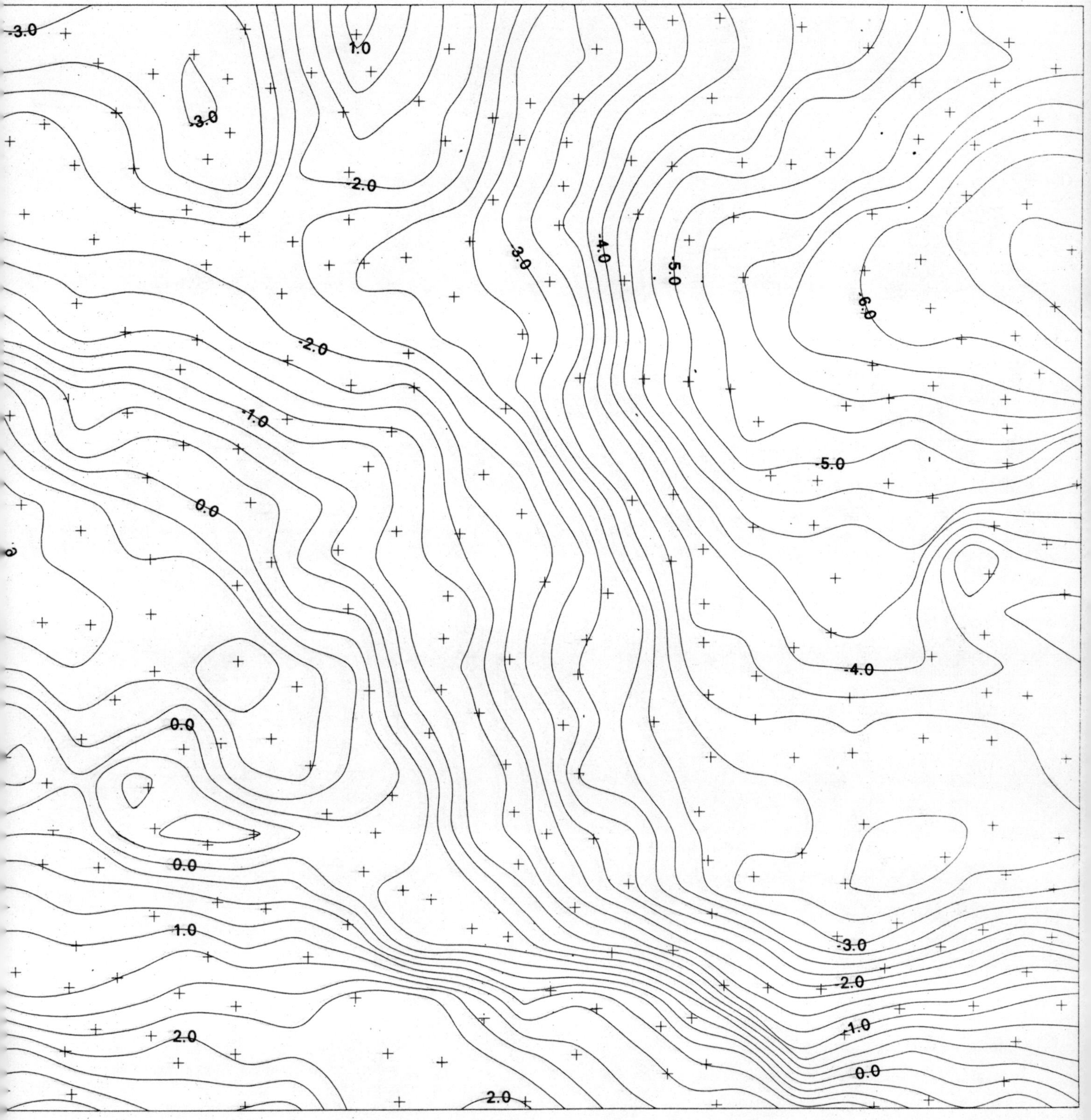

Map E.· Automatically generated contours of Bouguer Anomaly Values. (See page 30)

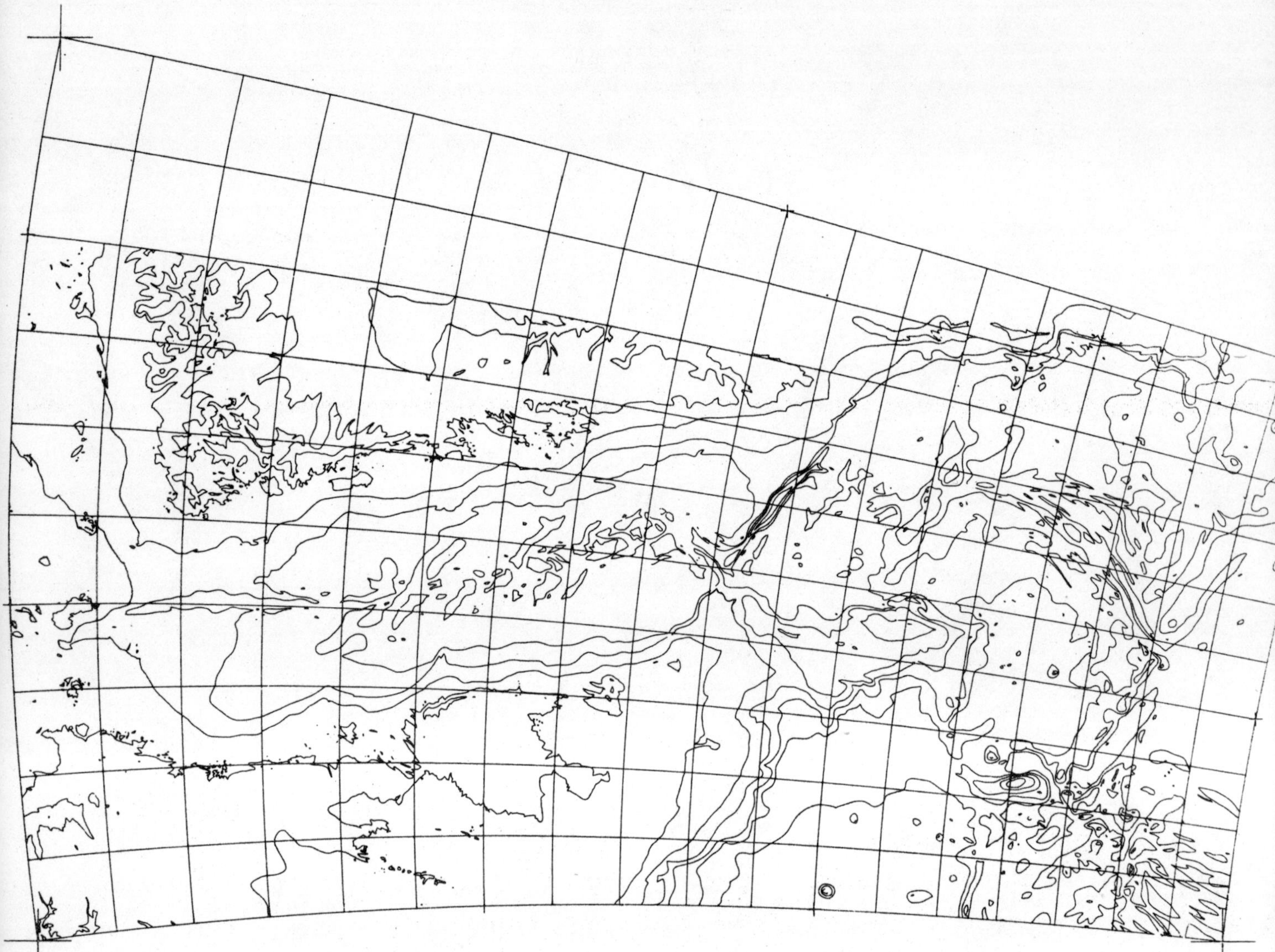

Map F. Gulf of Aden projection change

(See page 30)

Acknowledgements

The maps in this book were produced from data supplied by government institutions working with the ECU on joint projects. Maps A and F were derived from data from charts produced by the National Institute of Oceanography, whilst maps B and D were produced from digitised Ordnance Survey large scale maps and plans. Maps C and E were produced from data made available by the Institute of Geological Sciences.

Printed by Cook, Hammond & Kell, Ltd., Westminster and Mitcham.

5.7 Programs under development

This section describes briefly some of the programs which are still under development, but which are expected to come into general operation shortly. (The detailed operating instructions for these programs are not included in section 6.)

EMATCH is a program which will match the two ends of any line crossing the boundary separating two sheets of a mosaic, to ensure an accurate junction. The program operates on the combined mosaic of sheets as output by the AMALG program.

FCI, FAM and FMATCH are the first in a series of programs which operate with data dealing with "fields". In this context, a "field" is defined as a two-dimensional enclosed area having some unique property. (E.g. a map describing the soils of a district is composed of "fields" where each field corresponds to some patch of a specific soil type.) It is operationally convenient to associate with each field an "area peg", which is a point approximately at the visual centre of the field. The properties of the field are then stored as the feature codes of the corresponding area pegs.

FCI simplifies the digitising of field maps. When a boundary line is digitised, the area pegs of the fields to the right and left of the line are also digitised. This is much more convenient than entering numeric feature codes for the line. The codes are entered by typing off-line a paper tape, specifying one code for each field. FCI (Field and Crop Identification) merges these codes with boundary lines and area pegs.

FAM (Field Area Measurement) checks each field perimeter for continuity and for correct polarity of right and left area pegs. It then computes the area of each field and stores this as an additional feature code associated with the corresponding area peg.

FMATCH (Field Matching) operates on a mosaic of field maps, merging fields which cross map boundaries into single large fields, each with one area peg.

These programs will be followed by others, which permit:-

a) amalgamation of fields with similar properties into single larger fields;

b) selection of fields for drawing according to size (to facilitate generalisation at reduced scales);

c) automatic placement of symbols and labels at area peg positions.

6. Detailed Program Operational Instructions

This section consists of the detailed operating instructions for all the programs mentioned above. The programs described are:-

AMALG	Amalgamate map files to single large file.
ARCHIVE	Store binary figure field in a named magnetic tape file.
CORECT	Apply selective corrections to a map file.
CSDT	Copy files from 7-track magnetic tape to DEC tape.
CTOUR	Produce contours of a figure field.
FFIN	Convert an imported figure field to internal binary format.
FFOUT	Convert an internal binary figure field to exportable format.
FIGURE	Prepare a paper tape of parameters necessary for figure field generation.
GETLST	Convert point data from standard internal format to X, Y, Z format prior to figure field generation.
GRATGN	Generate a map graticule.
IBMIN	Convert 7-track IBM tape from standard IBM to DEC format.
IBMOUT	Convert 7-track IBM tape from DEC to standard IBM format.
LINLAB	Produce line labels on the light spot projector.
LPP	Produce a catalogue of line symbol combinations on the light spot projector.
LSTCOD	List on paper tape the feature codes of lines and points.
MERGE	Merge feature codes punched off line on paper tape with map features.
MTCPY	Copy and verify magnetic tapes.
MTEDIT	Edit feature codes by interaction with the user.
NEARNS	Evaluate the nearest neighbour statistic for point data sets.
NGEAG	Draw maps on the Geagraph (i.e. pen or scriber).
NGEAL	Draw maps on the Gealux (light spot projector).
NGEANA	Draw "3 dimensional" anaglyph maps on the Geagraph

NJOIN	Orient map relative to projection axes and correct distortion. (9 map projections.)
NMTTDT	Convert tape produced on the D-mac to internal ECU format.
NSD	Draw maps on the storage display.
NSJOIN	As NJOIN for maps in a rectangular or non-standard projection.
PROJCH	Change map projection (9 map projections).
READFF	List binary figure field on the teletype.
RETRIV	Restore binary figure field from named file.
SAMPLE	Accumulate statistics of line lengths and point densities.
SLOPES	(also WAVER, SHADOW) Compute a figure field from point data.
SPP	Produce a catalogue of point symbol combinations on the light spot projector.
STG	Convert a map file from internal ECU format to standard export format.
STNDFF	Convert filed binary figure fields to exportable format.
STNDRD	Convert a map file from standard exportable format to internal ECU format.
TIDY	Reorder map files on tape.
TTP	Add a valid end of tape marker to an incomplete tape.

Operational Instructions - AMALG

Program name AMALG

Function To read a multi file tape output from NJOIN or NSJOIN and amalgamate chosen files into one single file.

Written by D. Vicars

1. Introduction

When a map is broken down into a mosaic of smaller sheets for digitising, each small sheet gives rise to a separate file on magnetic tape. When drawing the whole map at a later stage (or processing the data in the computer - e.g. for a projection change) it is inefficient to have to deal with a large number of small files. The AMALG program has been written to compact the separate small data files into one large single file. It also provides the option to select only features with specific feature codes for amalgamation, and to omit the rest. The scale at which the map is stored can also be changed. Excess data points on the features to be amalgamated can be filtered out, using a filtering technique identical to that employed by the Geagraph drawing program.

When separate files are amalgamated their respective co-ordinates have to be adjusted so that the corresponding maps will lie in their correct relative positions - just as when they are juxtaposed by the NGEAG program. This juxtaposition is achieved by adding to the co-ordinates for each file the co-ordinates of the local origin (as set up by the NJOIN program). This means that only files with the same local origin can be amalgamated. The local origin for the output file is set to the SW corner of the amalgamated map.

Only the four extreme corner points of the amalgamated map will be present; all other internal corner points are suppressed. The program runs in two or more passes. In the first pass a list of files is entered, and the input tape is scanned. The details of each file are listed. When this is finished the program types "END OF PASS 1" and requests the filtering tolerance, scale and code selection. The tape is read and the selected data transferred to the output tape in the second pass. The operator then has the opportunity of repeating this pass with a different filtering tolerance and code selection, and also possibly with a reduced file selection.

2. Program setup

Set up the two IBM units as follows:

MT5 Input tape (write disabled)
MT6 Output tape (write enabled)

The program will first type its name and date of issue. It then proceeds to ask a series of questions, these are best illustrated by example.

Q1: ENTER 1ST FILE NUMBER
R1: 1.

If the reply is "Ø." then every file or the input tape will be selected and questions Q2 and Q3 skipped otherwise the file given is located on the input tape and its descriptive information typed. This information is used as a standard for comparison with any other files selected. The output file will have this same permanent heading data. The printout is followed by:

Q2: IS THIS OK
R2: YES or NO

If the answer is "NO" then question Q1 will be repeated. Otherwise the program reads in a list of files in any order, asking for each one with

Q3: NEXT FILE
R3: 2.

The list is terminated by typing a zero or negative value. The program then searches the input tape for each file number printing out its sheet reference number, and an error message if there is any significant difference between it and the first file. This is terminated with the comment "END OF PASS 1". The program now requests the data to update the file header.

Q4: IS THIS THE FIRST FILE ON O/P TAPE
R4: YES or NO

If the reply is "NO" Q5 is skipped.

Q5: ENTER TAPE NUMBER
R5: MT 34.

This must be supplied as MT or DT then the number.

Q6: ENTER DATE AS DD-MM-YY
R6: 28-Ø-7Ø

The date is given a few elementary checks and if any of these fail then "ERROR" will be typed and the request repeated.

Q7: ENTER YOUR INITIALS
R7: D.V.

Only the first 5 characters are considered.

Q8: ENTER RUN DESCRIPTION
R8: AMALG TEST RUN

This is used to identify the run. Only the first 25 characters are retained.

Q9: ENTER TOLERANCE
R9: Ø.Ø1

The tolerance (which should be a distance in millimetres) is used to filter points on lines. If a point is closer to its predecessor than the nominated tolerance it will not be copied to the output file. If an entire line is shorter than the tolerance it will be omitted altogether.

Q10:ENTER SCALE
R10:N Ø.1

The scale must be preceeded by a letter to qualify the number

A Absolute scale, followed by the scale e.g. A 25ØØØ. produces an output at 1/25ØØØ.

D Scale relating to digitised map e.g. D Ø.1. This makes allowance for differences between the nominal scale and the digitised scale.

N Nominal scale factor

S Fit to scope size. The scale factor is calculated so that the width of the resulting map will fit onto the storage display.

X) Y) Scale factors are calculated to fit the map to the given X and Y dimensions in centimetres.

The error message "PREFIX P UNKNOWN" is given if the prefix is not one of these. The absolute scale and scale factors are typed out for inspection and can be accepted or rejected by the reply to question 11:;

Q11:0/P SCALE = 25ØØØ.
SCALE FACTORS, X = Ø.1ØØØ Y = 0.1ØØØ
TOLERANCE = Ø.Ø1ØØ IS THIS OK

R11: YES or NO

If the reply is "NO" then the tolerance and scale must be entered again.

Questions 12 to 17 are asked by a standard subroutine "SELECT" which is used to obtain from the user a specification of which items within the sheet are to be drawn and which omitted.

Q12: ENTER CODE SELECTION
END WITH 99999.

R12: 32.
79.
-164.
99999.

As a reply to Q7 the user must specify a list of feature codes. Each code must be terminated by a decimal point and the whole list terminated by 99999. Items with these codes will be selected for playback.

If the list of codes is empty (i.e. only 99999 has been entered) it is assumed that all items on the sheet are to be drawn and questions Q14 to Q17 are omitted.

Q14: SEARCH FROM FEATURE CODE NO.
R14: 3.

Q15: TO FEATURE CODE NO.
R15: 5.

If each item on the tape had 8 features only codes 3, 4 and 5 of each item would be compared against the list of selected codes.

Q16: IS SELECTION TO BE INCLUSIVE
R16: YES

If the reply is YES then items with the nominated codes are copied to the output file. If the reply is NO then the list of codes is taken to be exclusive i.e. items with these codes are omitted and all the rest copied to the output file.

Q17: DO YOU WANT INTERNAL LINES SUPPRESSED
R17: YES

If the reply is YES, then items with more than one code appearing in the list of selected codes will be omitted, e.g. If codes for chalk and clay boundaries had been specified, boundaries between chalk and clay would be omitted.

At this stage the second pass of the input tape is made, and the features which have been selected are copied to the output file.

At the end of the pass the program types "END OF PASS 2" and asks question 18

Q18: DO YOU WANT ANOTHER SELECTION
R18: YES

This enables the second pass to be repeated with a reduced file selection, and a different tolerance and code selection. If the reply is "NO" the program will close the output tape and stop.

Q19: DO YOU WANT ANY FILES DELETED FROM LIST
R19: YES

If the reply is NO, question 20 is omitted.

Q20: ENTER FILE NUMBER
R20: 7.

The program replies either "FILE 7 NOT FOUND" or "FILE 7 DELETED" and repeats Q19. This procedure is terminated by entering a zero or negative file number.

Q21 to Q27 are the same as Q12 to Q17 and allow a new selection of features to be made.

Q28: ENTER TOLERANCE
R28: Ø.Ø5

The pass of the input files to select the new group of features is then repeated. The message "END OF PASS 3" will be typed when it has finished. The program then loops back to Q18.

Operational Instructions - ARCHIV

Program name ARCHIV

Function To store a figure field in a named file

Written by D.S. Connelly

1. Introduction

Figure field storage costs are reduced when many figure fields are stored together on one file-oriented I.B.M. tape. However, most of the figure field and contouring routines are designed to process non-file-oriented I.B.M. tapes only. ARCHIV will transfer a single figure field from a non-file-oriented tape and incorporate it into a file-oriented archive tape.

ARCHIV is not a general purpose storage routine. Only standard figure fields may be processed by it.

2. Program setup

Mount the input MAGTAPE on unit 5, ring removed. Mount the archive (file-oriented) MAGTAPE on unit 6. The DISC must be operating with DISC1 file-oriented, write-enabled and containing at least 200 free blocks.

Mount the contouring operations DECTAPE on unit 2. Initiate the DEC MONITOR and type E ARCHIV to fetch the program.

The program then requests the name which is to be given to the output file.

Q1. GIVE NAME OF OUTPUT FILE (6 LETTERS + BIN)
R1. CHALK2BIN

Type the file name as 6 letters and/or numbers and/or spaces followed immediately by the letters BIN.

If a file of this name already exists on the archive tape, that file may be deleted or the chosen name may be changed.

Q2. FILE CHALK2BIN ALREADY ON ARCHIVE TAPE. DO YOU WANT TO DELETE THIS FILE
R2. NO

Type YES to delete the current file of this name. If NO is typed, the output file must be renamed.

Q3. GIVE NEW NAME TO OUTPUT FILE
R3. CHALK3BIN

When the figure field has been written on the archive tape, the program will begin again with a request for an output file name. If another figure field is to be stored on the archive tape, mount the new input tape and carry on. If you are finished, type C to return to the MONITOR.

Operational Instructions - CORECT

Program name CORECT

Function To supply selective correction to a digitised map sheet

Written by T. Lang

1. Introduction

The EDIT program provides a convenient and thorough means of checking and correcting digitised map data, particularly where there are many corrections or the data is to be checked for the very first time. EDIT steps through the digitised sheet item by item, and uses the sequence in which the items are played back to indicate which item is to be corrected next.

Where only a few items out of many are to be corrected the EDIT approach is not so convenient. In this situation CORECT provides an alternative, in which only the items to be corrected need to be considered by the user. The items to be corrected are specified by "pointing" at them on the D-mac. As an example consider the case where a line is to be deleted. If a human editor was giving instructions to a draughtsman he would say "delete that line", and point with his finger to indicate exactly which line. Exactly the same instruction may be given to the CORECT program as was applicable for the draughtsman. In the case of CORECT the instructions are supplied via a D-mac tape. The editor types "DL" on the D-mac keyboard (signifying <u>D</u>elete <u>L</u>ine) and then digitises a point on or near the line he wishes to be deleted.

CORECT takes as data the old sheet and the correction sheet, and merges them to form the new updated sheet. First the correction tape is scanned, and a list is built up of those points where a correction is to be applied. Then the old sheet is also scanned. Each point (i.e. individual point or point within a line) is compared against the list of points requiring corrections. "Comparing" consists of comparing the x, 7 co-ordinate values. If a match is found, the corresponding correction is applied and the correction removed from the list of corrections. If there is no match the old data point is copied directly to the new data tape.

2. Preparing the correction tape

The correction tape is prepared on the D-mac exactly as for any normal data tape, except that every item must be prefixed by a two-letter correction code. As its name implies, the correction code specifies exactly what correction is to be applied; the co-ordinate data (individual point or line) which follows specifies where that correction is to be applied.

Table I Correction codes

Correction code	Followed by	Internal Code No.	Action required
DL	point	1	Delete the line which passes through the correction point
DP	"	2	Delete the point which occurs at the correction point
IL	line	3	Insert the new line which follows
IP	point	4	Insert the new point as specified
PL	"	5	Print out on the teletype the feature codes of the line which passes through the correction point
PP	"	6	Print out on the teletype the feature codes of the point which occurs at the correction point
CL	"	7	Change the feature codes of the line which passes through the correction point (the new codes are as specified for the correction point)
CP	"	8	Change the feature codes of the point which occurs at the correction point (the new codes are as specified for the correction point)

The codes IL, IP, CL, CP should be followed by feature codes as appropriate.

The correction codes are entered exactly as if they were feature codes i.e. the "KEYBOARD" button should be pressed, followed by the two keyboard keys corresponding to the correction code.

As an example, consider the preparation of a correction tape where the actions required are:

(a) to replace one line by a new version
(b) to change the feature codes of another line

The correction tape would be made up as follows:-

corner points	as for normal sheet
< K> DL	
point 1	point on line to be replaced
< K> IL	
< K>36	feature codes of replacement line
< K>71	
line 1	replacement line
< K> CL	
< K>84	new feature codes for 2nd line
< K>-92	
point 2	point on line which is to have its codes changed
end sheet	as for normal sheet

Correction codes PL and PP could be used to find out exactly what feature codes specified points or lines have. If these feature codes prove to be in error, then a further run of the CORECT program (with a new correction tape) will be needed to put them right.

The correction tape must be processed by NMTTDT and NJOIN before it can be used by CORECT. The operational instructions for NMTTDT describe how that program checks a correction tape to ensure that every item has a valid correction code. The processing by NJOIN is also essential, since only when the sheet co-ordinates have been rectified can a direct comparison be made between the co-ordinates as specified on the old map sheet and on the correction sheet.

3. How the program works

This section describes the internal actions of the program. This explanation is included to help the user

(a) understand what sort of data the program requires

(b) select the data judiciously so that the program can run quickly on the computer

(c) diagnose the sources of error if the program is unable to make all corrections successfully.

The program reads the old data sheet from one IBM tape unit, and writes the new updated sheet to the second IBM tape unit. As this utilises both IBM units, the correction data has to be read from a DEC tape unit. (It is a trivial operation to transfer the correction data from IBM tape to DEC tape before CORECT is run. A special program CSSDT - Copy a Single Sheet to DEC Tape - is provided to do this).

CORECT first reads the heading information from the sheet to be corrected, and from the sheet of correction data. This enables checks to be made that both sheets do exist on their respective tapes and that the correction sheet is in fact a correction sheet and not an ordinary data sheet. The output data tape is then positioned ready for writing the updated output sheet.

During the next phase the rest of the correction data is scanned. Any items marked as insertions (i.e. IL and IP) are copied direct to the output sheet. All the other items are entered into a table in core store. The table contains entries listing

(a) the type of correction required (e.g. DL)

(b) the co-ordinates of the point where the correction is to be applied. Corner points are both copied to the output tape and stored in core store.

The size of the correction table is limited by the size of the core store. The current size limit is 50 corrections and 200 associated feature codes. (It may be possible to raise these limits later). As the correction table is built up, the opportunity is taken to check that all corrections are correctly coded. This check should already have been performed when the correction sheet was processed by NMTTDT, but the second check is an additional safeguard. At the end of this stage the correction tape has been finished with, and is rewound.

The next phase is the correction phase proper. The old data sheet is scanned item by item, and the co-ordinates of each item compared against the list of co-ordinates stored in the table of corrections.

First of all the corner points of the old sheet and the correction sheet are compared. If there is any significant difference in position between the sheets a warning message is printed. This difference could mean that the correction sheet is being matched to the wrong data sheet; if this is the case the user may abandon the program at this point. Because the corner points of the correction sheet have already been copied to the output tape, the corner points of the old data sheet can be discarded once they have been checked. (Advantage may be taken of this exchange of corner points to update the corner points of the old sheet, should this be necessary).

Individual data points (as opposed to points on a continuous line) are compared against the co-ordinates stored in the correction table. If no match is found the old point is copied direct to the new sheet.

If a match is found the appropriate corrective action is taken (e.g. DP, CP, PP). The correction table entry is then deleted (to speed up the checking of all following data).

It is unlikely that there would be a perfect match between the co-ordinates of points on the correction sheet and old data sheet. To overcome this problem the user is allowed to nominate a "threshold". If the x, y co-ordinates compare to within the limits of this threshold a match is assumed. The effect of the threshold is to place a square box of side (2 x threshold) centred on the correction point. If the old data point falls within this box a match is assumed - see fig. 1.

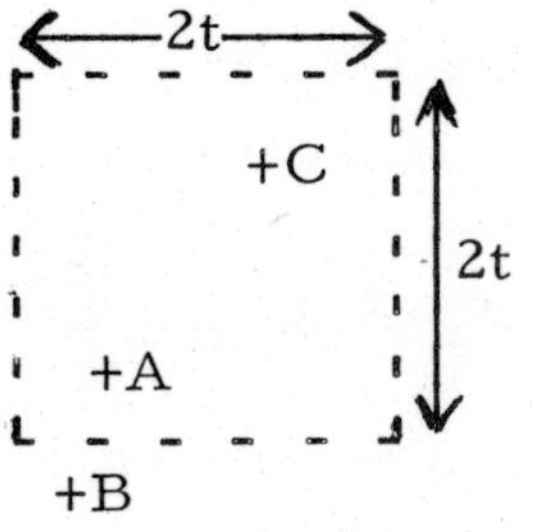

C = correction point
A = matching data point
B = non-matching data point
t = threshold distance

Figure 1 - Use of "threshold" to determine point match

It is up to the user to specify a threshold distance as large as may be necessary to nullify the effects of errors in the digitising system. But if the threshold is too large, a match may be found with some other point than the point intended. The program applies the correction to the first matching point found. Because the correction entry is deleted from the correction table at this time, there can be no warning if a second point is found which would also have given rise to a match.

In most cases line data consists of a string of closely separated points (e.g. 0.25 mms. apart at the scale of digitising). In these circumstances, it is possible to treat each line point as an individual point, except that if a match with a correction point is found the correction is applied to the whole line rather than just to the point which caused the match.

The comparison of every line point against every correction table entry is a time consuming task, even when fast integer arithmetic is used. However, if a fairly large threshold has been specified then it is reasonable to assume that on any one line several data points will fall within the threshold box surrounding a correction point. To reduce program running time, the user is therefore asked to specify at what frequency line points are to be compared with the correction points. If the user sets the "line by-pass count" to 1, then every point on a line will be checked, if 2 then only every 2nd point will be checked and so on.

Thus the CORECT program will run most quickly when both the threshold and line by-pass count are as large as possible. But if the threshold is to be large, then there is a danger that the correction point may be found to match some line other than the line for which it was intended. This danger can be minimised by choosing correction points with discretion. Figure 2 illustrates this. Suppose that line A-B is to be deleted. A good position to digitise the correction DL is point p, and not points q, r, or s, which are close to other neighbouring lines.

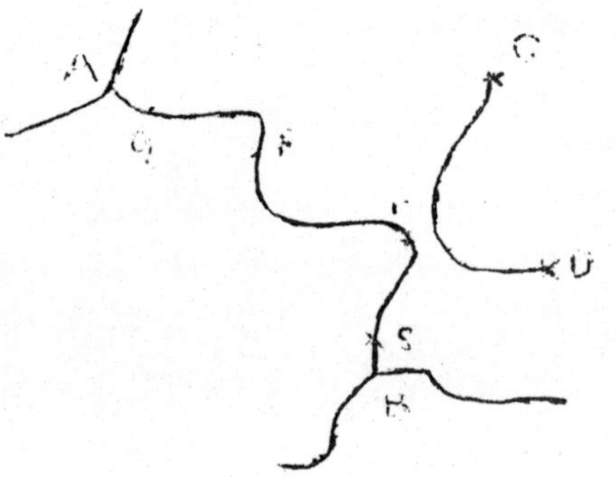

p is good
q, r, s are too close to other lines.

Figure 2 - Selection of a correction point for a line

In some cases however, successive points on a line are not close together. This occurs where the digitiser operator has noticed a long section of straight line and moved the digitising pencil directly from one end of the line to the other without recording any intermediate points. In these circumstances it is no longer sufficient merely to check each line point to see if it lies near a correction point. A check must also be made whether the straight line connecting the line points passes close to a correction point, as illustrated in Figure 3.

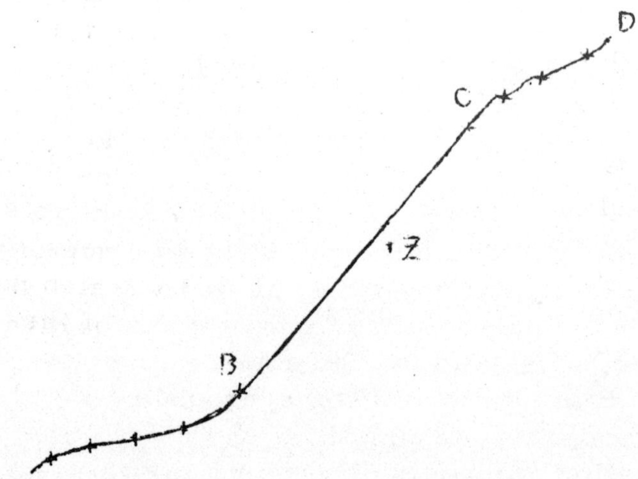

Figure 3 - Correction of a line with widely spaced points

Therefore CORECT checks the spacing of successive points on a line. If the points lie closer together than the threshold distance, then the normal point by point comparison with correction points is employed, as described above. But if the point separation is greater than the threshold distance then interpolation is employed to see if the connecting line passes close to a correction point. It would be possible to compute whether the connecting line intersected the box round the correction point, much as when closely spaced points are processed. However, in order to reduce computing time, the box is replaced by a cross centred at the correction point. If the connecting line intersects the cross a match is said to have been found with the correction point - Figure 4.

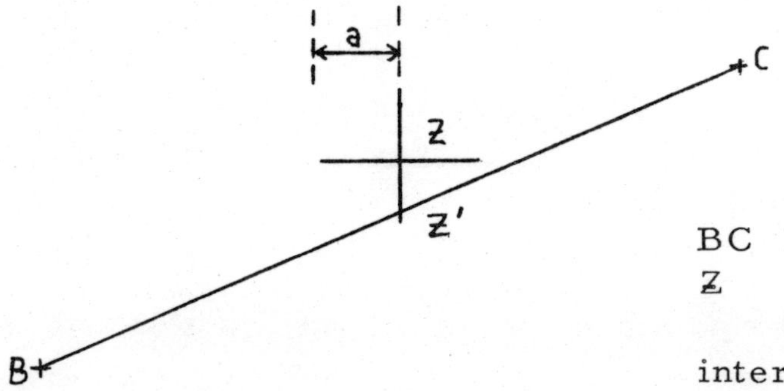

BC line to be corrected
Z correction point

intersection at Z' indicates that a "match" has been found

Figure 4 - Checking widely spaced line points for match with a correction point

It could be the case that a line which would intersect the standard box round a correction point would not intersect the cross. (Particularly a line at 45° to the map axes). To overcome this difficulty the overall dimension of the cross has been made slightly larger than that of box. The length of the arm of the cross (distance (a) in figure 4) has been chosen such that $a = 1.6t$ where t is the threshold distance.

Once the old data tape has been scanned the end marker can be planted in the newly updated sheet, and both old and new data tapes are then rewound. As a last step the program checks to see that the correction table kept in core is empty. Any entries which are left signify corrections on the correction tape for which no matching items were found on the old data tape. Details of these entries are printed on the teletype.

4. Program set up

The program starts by typing its title, followed by the instructions

LOAD CORRECTION DATA ON DT4
INPUT TAPE ON MT5
OUTPUT TAPE ON MT6
PAUSE

The tape units should be loaded as directed. Only the output tape should be write enabled. Type ↑P on the teletype to continue from this point.

(If the program discovers later that one of the tape units is not set up correctly the warning

IOPS4

will be given. Reset the unit correctly, and type ↑R to continue.)

The program goes on to ask a series of questions -

Q1. ENTER NUMBER OF SHEET TO BE CORRECTED
R1. 5.

The program searches for the nominated sheet on the input data tape. If the sheet can not be found

SHEET NOT FOUND

will be typed, and Q1 repeated.

Q2. ENTER SHEET NUMBER ON CORRECTION TAPE
R2. 1.

The reply will usually be 1, as there will rarely be more than 1 sheet on the correction DECtape. The program will check that this sheet is in fact a correction sheet. If it is not, the message

NOT A CORRECTION SHEET
PAUSE

will be typed. In this event the user should type ↑C to abandon the program or ↑P to continue.

Q3. ENTER THRESHOLD IN MMS
R3. 5.5

This is the threshold distance used in the comparison of point co-ordinates, as described in section 3.

Q4. ENTER COMPARISON COUNT (1.0 = COMPARE EVERY POINT)
R4. 4.

This is the "line by-pass count" also described in section 3.

Q5. IS THIS THE FIRST FILE ON O.P. TAPE
R5. NO

If the reply is NO, the output tape is spaced forward past the information written at some earlier time. If the reply is YES the question is asked -

Q6. ENTER TAPE NUMBER
R6. MT71.

The reply for this program should always be MT, followed by the tape serial number.

Questions Q7 to Q9 are used to update the header information in the new output sheet.

Q7. ENTER DATE AS DD-MM-YY
R7. 03-11-69

Q8. ENTER YOUR INITIALS
R8. AJK

A maximum of 5 characters may be used.

Q9. ENTER RUN DESCRIPTION
R9. TO ERASE ALL WELLS

This is a temporary title used to label only the new version of the output sheet. It will generally be left blank. A maximum of 25 characters may be used.

At this point the new updated sheet is started on the output tape. If the program is abandoned after this stage, special steps will be required (e.g. use of the program TTP) to erase the new sheet and plant a new end marker on the tape.

The program then scans the correction sheet. The data is checked for consistency. A number of error messages may be given -

INVALID MARK NUMBER 9 ON CORRECTION SHEET
PAUSE

The mark number should indicate the type of item read (e.g. individual point, start of line, etc.) Type ↑P to ignore the faulty item and continue, ↑C to abandon the program.

INVALID CORRECTION CODE 19 ASSOCIATED WITH POINT
PAUSE

The correction codes are the numbers used inside the computer to represent the corrections DP, IP, etc. The list of legal code numbers is given in Table I. Type ↑P to ignore the item, ↑C to abandon the program.

INVALID CORRECTION CODE 0 ASSOCIATED WITH LINE
PAUSE

Again type ↑P to ignore the item, ↑C to abandon the program

INVALID MARK NO. 4 IN SERTION LINE
PAUSE

If the user chooses to continue, the line to be inserted on the updated sheet is terminated at this point.

NUMBER OF EDIT CORRECTIONS EXCEEDS 50
STOP

"Edit" corrections include all corrections except insertions. The program is forcibly abandoned (but with the partially formed output sheet correctly terminated).

NUMBER OF FEATURE CODES EXCEEDS 200
STOP

This is the maximum total number of feature codes which may be associated with "edit" corrections. Again the program is forcibly abandoned.

Once the correction sheet has been successfully scanned, the correction tape is rewound. The correction phase proper is then started. The first four corner points from the correction sheet are compared against the corner points on the input sheet which is to be corrected. If there is a positional difference between the two sheets greater than the threshold distance a warning will be typed

CORNER POINT 2 MISS-MATCH BY 0.74 CMS

The program will continue running, but the user may type ↑C to abandon it if he so chooses.

The only other messages arising from this phase of the program will be those generated by the pseudo-corrections PL and PP -

LINE THRU 17.43 16.10 HAS 3 FEATURE CODES
9001 -71 10004

and

POINT AT 6.15   18.38 HAS 1 FEATURE CODES
49
(The co-ordinates are given in centimetres).

When this phase is completed both the input and output tapes are rewound. The program then checks to see if it has any corrections left unsatisfied. If there are none, the message given is

ALL CORRECTIONS MADE

If corrections are left they are listed, complete with their X, Y co-ordinates.

2 EDIT DIRECTIVES REMAIN UNSATISFIED

X	Y	EDIT
19.23	0.14	DL
77.01	14.97	IP

The program terminates with the final message

FINISHED
STOP

Operational Instructions - CSDT

Program name CSDT

Function To copy map sheets from magnetic tape to DEC tape

Written by T. Lang

1. Introduction

This program allows the user to copy several map sheets from magnetic tape to DEC tape. The copy process is a direct copy from one tape to another. The sheet header blocks are not updated to indicate that the new sheets are stored on a different tape.

2. Program setup

The program starts by typing the message

PROGRAM TO COPY SHEETS FROM MT5 TO DT4

PAUSE

The tape units should be loaded as indicated, with MT5 write-disabled, DT4 write-enabled. Type ↑P to continue.

The program then asks

Q1. ENTER MT SHEET NO.

R1. 3

The program will search for sheet 3 and copy it to DEC tape. If the specified sheet cannot be found on the source tape, a message will be printed

SHEET NOT FOUND

and Q1 repeated.

When the sheet is successfully transferred, the program goes on to ask

Q2. ANY MORE SHEETS FROM THIS TAPE

R2. YES

If the reply is YES the program loops back to Q1. If the reply is NO, the program asks

Q3. ANY MORE TAPES

R3. YES

If the reply is YES, then the program prints the instruction

LOAD NEW TAPE

PAUSE

MT5 should be loaded with the new tape. Type ↑P to continue; the program will loop back to Q1.

If R3 is NO, the program will add an end of tape marker to the DEC tape, and terminate with the message

FINISHED

STOP

Operational Instructions - CTOUR

Program name CTOUR1, CTOUR2, CTOUR3

Function To produce countours of a figure field

Written by D.S. Connelly

1. Introduction

There are three separate programs available for contouring. They differ only in their speed and their flexibility for input and output. They do not produce different interpretations of the same figure fields.

CTOUR1: This is the standard production contouring program. It is designed to use the disc as efficiently as possible. However, should the disc be unavailable or filled with protected information, CTOUR1 cannot be used. CTOUR1 will allow output to DECTAPE or MAGTAPE in the standard internal E.C.U. format or to the CRT storage display. Direct output to the Geagraph is not allowed, but a stored map may be readily played back on the Geagraph when the contouring is complete. The figure field input may be from either file-oriented or non-file-oriented MAGTAPE.

CTOUR2: This program provides a back-up system to CTOUR1. It is capable of working without the disc and with only one MAGTAPE drive; it would, however, operate rather slowly in this mode. A large degree of flexibility is afforded the user for structuring the peripheral configuration for maximum speed.

CTOUR3: This is the "quick-look" program. It uses the disc and the CRT and will not work without either of these devices. It produces somewhat angular contour lines on the CRT which are of sufficient quality for a quick look at the shape of the figure field and the contour lines produced by this system. CTOUR3 will process a 50 x 50 figure field in under 5 minutes, each contour elevation taking about 30 seconds.

2. Facilities and restrictions

(i) Translation of origin: CTOUR1 and CTOUR2 allow the user to specify the co-ordinates of the lower left (SW) corner point on all output devices except the CRT. For CRT output in these programs and in CTOUR3 the image is automatically positioned for optimum visibility.

(ii) Scale change: CTOUR1 and CTOUR2 allow the user to specify a scale factor relative to the input figure field. Separate X and Y scales are not available. There is no facility for specifying the scale cartographically (1 inch = 1 mile). The supplied scale factor is applied to all output except to CRT output. CRT output in these programs and in CTOUR3 is automatically scaled so as to fit the largest possible map completely on the CRT face.

(iii) Review of parameters: All three programs allow the user to view and approve the figure field parameters and the contouring elevation parameters.

(iv) Addition of extra contours: All three programs allow the user to specify extra contours to be added once the main contouring task is complete.

(v) Diagnostics: CTOUR1 and CTOUR2 have complete, built-in diagnostics. CTOUR3 merely notes errors and proceeds to the next contouring elevation.

(vi) Input format: Figure fields presented to all three programs must be those produced by the E.C.U. figure field generating programs or have a format identical to those figure fields. All three programs accept non-file-oriented figure field storage. Only CTOUR1 and CTOUR3 accept file-oriented figure field storage as well.

(vii) There are no facilities for labelling maps or contours in the graphic output. Hachures are not attached to depression contours. All contours within the contouring frame are shown, including unreliable contours and closely spaced contours.

(viii) Figure field size: All three routines are capable of accepting up to 1850 contour positions lying on grid lines for each contour elevation. For each contour elevation there may be 100 disjoint contour lines. Because of these limitations it is not advisable to contour a figure field which is denser than 120 x 120. However, the program is capable of accepting figure fields which are 201 x 201. The simplest method of deciding if a dense figure field can be contoured with these programs is to try it out with no output device selected.

(ix) Mesh size: In all three programs the mesh size at the output scale must be larger than 0.1mm in both dimensions.

(x) Progress reports: All three programs print the elevation of the current contour unless this printing is suppressed with switch 16. CTOUR1 and CTOUR2 also print the number of contour points found at each elevation. This print out may also be suppressed with switch 16. The larger is the number of contour points found the more laborious are the subsequent computations and, hence, the longer is the running time.

(xi) Switch settings: The console accumulator switches (right hand set of switches, numbered 0-17) provide a convenient means for interacting with the program as it is running. Their use also eliminates much slow teletype interrogation. For normal operations the switches are set at the beginning of a run and then left set until the run is complete. The switch settings have approximately the same meanings in all three programs and are summarised in the following table.

2.1 Summary of Switch Settings

Switch	Action if switch is up
Ø	Contour elevations are progressively incremented by the contour interval.
17	Contour elevations are progressively decremented by the contour interval. If switch 0 and switch 17 are both up, the current elevation is repeated. If switch 0 and switch 17 are both down, the program will PAUSE. The switches may then be reset. If both switches are left down and ↑P is typed after PAUSE, the current elevation is repeated.
1.	Output to Geagraph (CTOUR2 only).
2.	Output to CRT storage display. In CTOUR3 the down position of this switch does not inhibit CRT output altogether. In this program dots are displayed at the positions where contour lines cross the scan lines regardless of the setting of switch 2.
3.	Output to DECTAPE 6 in standard, internal E.C.U. format. A scratch device is required and DISC1 is assumed if none is explicitly selected. (CTOUR1 and CTOUR2 only).
4.	DECTAPE4 used for working space (scratch) in non-file-oriented mode. Previous data on this tape is lost. (CTOUR2 only).
5.	Output to MAGTAPE6 in standard, internal ECU format. A scratch device is needed and DISC1 is assumed if none is explicitly selected. (CTOUR1 and CTOUR2 only).

6. Not used.

7. DISC1 is used for working space (scratch) in file-oriented mode. The scratch data is stored in file SCRTCH BIN and other file-oriented data already on the DISC will be retained.

Simultaneous output to both scratch devices is allowed in CTOUR2, but only DISC1 will be used for input if both are selected.

When no output devices and no scratch devices are nominated, there will be no output and the computation of the splines will be suppressed.

8. Before reading the figure field for the next contour elevation, the program will examine the scratch device and write the contour lines stored there onto the selected output devices. (CTOUR1 and CTOUR2 only).

9. Save previous information on the output device. In the case of CRT output the storage display is not erased before the map output begins and the old map and the new map will be superimposed. This switch has no significance for Geagraph output. When output is to DECTAPE or MAGTAPE, setting this switch causes standard header information to be requested. In all cases this switch may be used even when there is no previous information on the output device.

1ø. Not used.

11. Not used.

12. Not used.

13. Not used.

14. Enable dynamic rewind. This switch must be down unless the program is being fetched from MAGTAPE (CTOUR2 only).

15. Abort. Processing is abandoned at the completion of the current task and the program restarts. However, during initial data input (program BEGIND) switch 15 has a different function. It causes a print out of the figure field parameters just read from the figure field tape. After printing these parameters

the program will PAUSE, ↑T causes a re-read. ↑P causes the program to continue. Remember to set switch 15 down after verifying the parameters.

16. Suppress print out. Non-essential teletype messages are suppressed to save time. During diagnostic print outs, this switch will dynamically suppress the remainder of a long list.

17. Causes the contour elevation to be decremented. See above.

3. Program residence

These three programs are in executable form and must be fetched from an on-line, file-oriented device. Either DISC, MAGTAPE or DECTAPE may be used, in that order of preference. In all cases the "C" handler must be used to fetch the program. That is, either DKC or MTC or DTC must be assigned to .DAT slot -4.

3.1 Normal operations

CTOUR 1: This program is best run under the DISC Working System. In cases where rather large figure fields are to be contoured, DISC space will be at a premium and the DECTAPE Working System must be used. In this situation the program must reside on DISCØ and the DISC Working System may be used. Under 'normal' (50 x 50 figure fields) circumstances, however, the program should reside on DISC1. The following steps should be followed to place the program on DISC1.

1. Mount the Contouring Operations DECTAPE on drive 1.
2. Initiate the DEC MONITOR (DISC Working System)
3. Type: A DTA1 2/DKA1 3/NONE 4, 5, 6, 7, 1Ø
4. Type: PIP
5. Type: N DK1
6. Type: T DK1 CTOUR1 XCT ← DT1 CTOUR1 XCT (B)
7. Type: ↑C to return to the MONITOR

To call the program, follow these next two steps.

1. Type: A DKC1 -4
2. Type: E CTOUR1

The MAGTAPE containing the figure field must be on drive 5. For normal operations the output will be directed to another MAGTAPE on drive 6. This tape should be mounted with the write-permit ring in place. Set switches Ø and 5 up. All other switches should be set down. (For CRT output switch 2 must be up as well). The co-ordinates of the lower left corner will usually be X=Ø.Ø. The scale factor will usually be 1.Ø. The border should be stored on the output tape.

CTOUR2: This program may very well be fetched from DISCØ in a manner similar to that described above for fetching CTOUR1 from DISC1. However, this program will usually be used when the DISC is unavailable. In that case the program should be fetched from MAGTAPE. First load the program on a file-oriented MAGTAPE in the following steps.

1. Mount the Contouring Operations DECTAPE on drive 1.
2. Mount a free MAGTAPE on drive Ø.
3. Initiate the DEC MONITOR (DECTAPE Working System)
4. Type: A DTA1 2/MTAØ 3/NONE 4, 5, 6, 7, 1Ø
5. Type: PIP
6. Type: N MTØ
7. Type: T MTØ CTOUR1 XCT ← DT1 CTOUR1 XCT (B)
8. Type: ↑ C to return to the DEC MONITOR

The next two steps will call the program.

1. Type: A MTAØ -4
2. Type: E CTOUR2

The output should be directed to DECTAPE6. Also, DECTAPE4 should be used as the scratch device. Set switches Ø, 3 and 4 up. All the rest should be down with the possible exception of switch 2.

CTOUR3: The program should be placed on DISC1 and then fetched in a manner directly analogous to that described above for CTOUR1. Switches Ø and 2 should be up. All other switches should be down. Remember to turn the CRT on before fetching the program.

4. Program setup

After the program has been fetched an introductory heading will be printed and operator response will be requested.

Q1. (CTOUR1 and CTOUR3)
THE IBM TAPE CONTAINING THE FIGURE FIELD MUST BE ON TU-20 UNIT 5 (RING REMOVED) (Suppress with switch 16)
IS THIS TAPE FILE-ORIENTED
>

R1. NO. Mount the figure field tape on MAGTAPE unit 5 and type YES or NO depending on whether the tape is or is not file-oriented. Type ↑T to restart the program. If you wish verification of the figure field parameters which are about to be read, set switch 15 up before answering the above query. Remember to set switch 15 down after the parameters have been verified.

Q1 (CTOUR2): WHEN THIS TAPE IS READY, TYPE CONTROL P.

PAUSE ØØØØØØ

R1: Mount the figure field tape on MAGTAPE unit 5. This tape must be non-file-oriented. Type ↑P when it is ready.

Q2 (CTOUR1 and CTOUR2): GIVE X CO-ORD OF FIRST CORNER FOR OUTPUT

R2: 100.0

Type the table X co-ordinate in millimetres of the first (SW) corner point, the point with minimum X and Y co-ordinates. You may return to this request by typing ↑T at any time during the data input segment of the program.

Q3 (CTOUR1 and CTOUR2):
GIVE Y CO-ORD OF FIRST CORNER FOR OUTPUT

R3: 100.0

Type the table Y co-ordinate in millimetres of the first (SW) corner point.

Q4 (CTOUR1 and CTOUR2)
GIVE THE SCALE FACTOR (NO PREFIX)

R4: 1.0

Type a single number giving the ratio between the output scale and the input scale. If this number is greater than one, the output map will be larger than the input map. Similarly, a scale factor less than one contracts the output size. Separate X and Y scales are not available.

Q5: CONTOURING FROM Z=100.0 TO 500.0 CI=100.0
CHORD LENGTH FOR OUTPUT IS 0.5MM
DO YOU WISH TO CHANGE THESE PARAMETERS

R5: NO

In this example contouring proceeds from elevation 100.0 to elevation 500.0 in steps of 100.0. The straight line segments approximating the smooth curves will have length 0.5 millimetres on the Geagraph. If these parameters are all right type NO. If you wish to change something type YES and answer the following questions.

Q5.1 GIVE LOWEST Z VALUE

R5.1 100.0

Type the elevation of the lowest contour to be plotted

Q5.2 GIVE HIGHEST Z VALUE
R5.2 500.0
Type the elevation of the highest contour to be plotted

Q5.3 GIVE CONTOUR INTERVAL
R5.3 100.0
Type the height interval between contour points

Q5.4 (CTOUR1 and CTOUR2)
GIVE CHORD LENGTH
R5.4 0.5
Type a chord length suitable for Geagraph output. Bear in mind any contemplated scale change at plotting time. (CTOUR2 - messages for CTOUR1 and CTOUR3 are similar)

Q6: SET SWITCHES
OUTPUT - 1=GEAGRAPH, 2=CRT, 3=DECTAPE6, 4=DECTAPE 4 (SCRATCH), 5=MAGTAPE6, 7=DISC(SCRATCH).
CONTROL - Ø = FORWARD, 9=JOIN, 14=REWIND, 15=ABORT, 16=NO PRINT, 17=BACKWARD
THEN TYPE CONTROL P

R6: Set the appropriate switches for output and ready the devices. If the Geagraph is being used it will require direct supervision. When the bell rings push the START button on the Geagraph console. This may have to be done many times during a run. For normal operations switch Ø should be set up. Switches 15 and 17 should be down. Switch 16 should be up only if the operator does not wish to note the program's progress through the contouring elevations. Once these and the output switches are set, type ↑P. The program will then leave the first (data-input) segment and the contour computations will commence. To repeat the data input segment, type ↑T instead of ↑P. If the output is being joined to existing data, switch 9 should be up. Questions necessary for joining will be asked. Refer to the Operational Instructions for NSJOIN.

The program should now cycle through the required contour heights. During each cycle it will type these messages.

M1. Z=100.0 (Suppress with switch 16)

The program is now computing the contour lines for the elevation 100.0. The switches directing the output should be reset now if they are not already correct. Until the figure field is completely read (until M2 is typed), the program may be advanced to the next contour by typing ↑T. Typing ↑T may also be used to make the program jump back to the previous contour elevation or simply to begin again according to the setting of switches Ø and 17. When in doubt, set all switches down and type ↑T. This will cause the program to PAUSE. Set switches and type ↑P to continue.

M2. (CTOUR1 and CTOUR2) 256 PTS.

(Suppress with switch 16). 256 contour points have been found on the figure field. The number of points is an indication of the length of time it will take to compute the contour lines at the current elevation. ↑T cannot now be used to restart the figure field reading.

At the completion of all the requested contours, the program will request action for finishing up.

Q7: GIVE 1 TO REPEAT OR ADD A SINGLE CONTOUR
GIVE Ø TO REPEAT OR ADD A SERIES OF CONTOURS
GIVE -1 TO FINISH UP

R7: Type 1 or Ø if more contours are required. The appropriate questions regarding the desired contour elevations will be asked. If -1 is typed, the program will then complete the contour map by drawing or storing the corner point marks and/or the bounding lines if they are desired.

When all tasks are complete the program will type

M3. -JOB ALL DONE-
TIME TAKEN Ø MIN, Ø SEC

STOP

The timer is not implemented for chained programs and the time is not properly tabulated. All work is now complete. Dismount all tapes, collect hard-copy output and tidy up the Geagraph, CRT, etc. Should switch 15 be up at this point, the program will not STOP but will restart with the request for fresh data.

Operational Instructions - FFIN

Program name	FFIN
Function	Converts a character figure field to a binary figure field.
Written by	D.S. Connelly

1. Introduction

The program performs the inverse of FFOUT. It reads a PDP-9 ASCII figure field and writes a standard PDP-9 binary figure field.

2. Program setup

Mount the PDP-9 ASCII figure field tape on MAGTAPE drive 5 with the ring removed. Mount a clean IBM tape with the ring in place on MAGTAPE drive 6. This is the output tape.

Mount the figure field operations tape on DECTAPE drive 2. Initiate the DEC MONITOR and type E FFIN.

The remainder of the process is automatic and takes one to five minutes. On completion the program will STOP.

Operational Instructions - FFOUT

Program name	FFOUT
Function	Convert a binary figure field to a character figure field.
Written by	D.S. Connelly

1. Introduction

The program (Figure Field OUTput) reads a PDP-9 binary figure field and writes PDP-9 ASCII figure field in the standard character format. This tape may then be processed by IBMOUT to produce a machine independent figure field.

2. Program setup

Mount the PDP-9 binary figure field on MAGTAPE drive 5 with the ring removed. Mount a clean IBM tape with the ring in place on MAGTAPE drive 6. This is the output tape.

Mount the figure field operations tape on DECTAPE drive 2. Initiate the DEC MONITOR and type E FFOUT.

The remainder of the process is automatic and takes one to five minutes. On completion the program will STOP.

Operational Instructions - FIGURE

Program name FIGURE

Function To prepare a paper tape of parameters necessary for figure field generation

Written by D.S. Connelly

1. Introduction

There are at present three programs, SLOPES, WAVER and SHADOW which produce figure fields from a set of data points. They are described in detail as a group, under the heading of SLOPES. Each of these programs requires as input a paper tape of data parameters and a DECTAPE file of data points in 3F12.5 format.

The data parameters may be prepared off line or with the aid of FIGURE. Both methods are described in what follows.

2. Program setup

The DEC MONITOR should be initiated. FIGURE should then be fetched from DECTAPE2. Mount the contouring operations DECTAPE on drive 2. Type E FIGURE.

After an introductory heading the program requests information from the operator. All numbers typed in response to requests from this program must contain a decimal point and should not occupy more than ten spaces exclusive of rubouts and deleted characters. A complete line may be deleted by typing ↑U before terminating carriage return is given. After the carriage return has been typed on an erroneous line, the line may be corrected only by restarting the program by typing ↑T. ↑T may be typed at any time.

Some of the quantities requested by the program are shown in the following diagram.

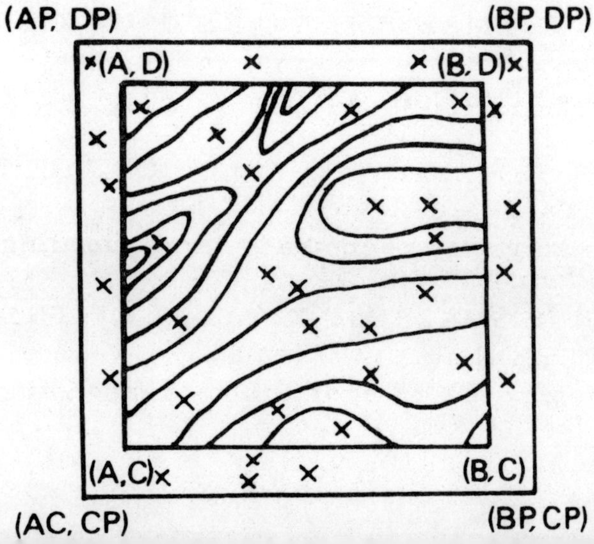

The program asks a series of questions about the data points and the figure field.

1. Request: GIVE NAME OF DATA POINT FILE (6 LETTERS + SRC)

   Response: Type the name of the DECTAPE file which contains the data points as X, Y, Z lists in character form. This name must consist of 6 letters and/or numbers and/or spaces followed by the three letters "SRC", 9 characters in all.

2. Request: GIVE NUMBER OF DATA POINTS

   Response: Type the number of points in the X, Y, Z DECTAPE file. Remember to terminate this number with a decimal point. This number must be correct, so double check it. If the data point list was obtained from the standard E.C.U. tape storage with the program GETLST, the number of points in the list is given in the output of that program. If this number is less than 4 or greater than 3000, the data point list cannot be processed by this program.

2.1 Advice: ERROR** NUMBER OF DATA POINTS = (NUM)

   Response: The program now returns to Request 2. Either type the correct number or abandon the program.

3. Request: GIVE SMALLEST X-VALUE OF THE PLOTTING FRAME

   Response: Supply the co-ordinate (in millimetres) corresponding to AP. It should be consistent with the co-ordinates used in the X, Y, Z data point list. These co-ordinates will be treated as the absolute display co-ordinates. However, all the display routines allow translation of the output frame as well as scale change during output.

4. Request: GIVE LARGEST X-VALUE OF THE PLOTTING FRAME

   Response: Supply the co-ordinate corresponding to BP.

5. Request: GIVE SMALLEST Y-VALUE OF THE PLOTTING FRAME

   Response: Supply the co-ordinate corresponding to CP.

6. Request: GIVE LARGEST Y-VALUE OF THE PLOTTING FRAME

   Response: Supply the co-ordinate corresponding to DP.

7. Request: GIVE SMALLEST X-VALUE OF THE CONTOURING REGION

   Response: Supply the co-ordinate corresponding to A.

8. Request: GIVE LARGEST X-VALUE OF THE CONTOURING REGION

   Response: Supply the co-ordinate corresponding to B.

9. Request: GIVE SMALLEST Y-VALUE OF THE CONTOURING REGION

   Response: Supply the co-ordinate corresponding to C.

10. Request: GIVE LARGEST Y-VALUE OF THE CONTOURING REGION

    Response: Supply the co-ordinate corresponding to D.

11. Request: GIVE NUMBER OF INCREMENTS PARALLEL TO X AXIS.

    Response: Supply the number of segments into which the figure field grid divides the edge of the contouring region which runs parallel to the X axis. This number should be an integer, but it must be terminated with a decimal point. Example (figure below): 16.

12. Request: GIVE NUMBER OF INCREMENTS PARALLEL TO Y AXIS

    Response: Supply the number of segments into which the figure field grid divides the edge of the contouring region which runs parallel to the Y axis. This number should be an integer, but it must be terminated with a decimal point. Example (figure below): 16.

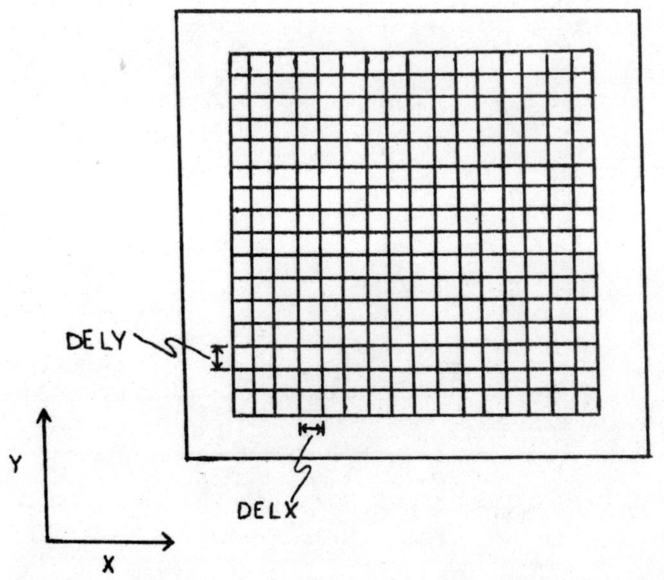

13. Advice: GRID MESH SIZE (DELX) X (DELY)

    Response: These are the dimensions of the cells of the figure field mesh. If either number is less than 0.1, there may be errors in the figure field and in the contours drawn from it. If either of these numbers is larger than 50.0, it is very likely that an error has been made on data input.

14. Request: GIVE FACTOR FOR Z CONVERSION ACTUALZ=DMACZ * FACTOR.

Response: Supply a factor such that when the data point Z values are divided by it and then truncated to integers, they will retain all their significant digits. In most cases this is the same number which was supplied to scale the feature code Z values during the generation of the X, Y, Z data point list by the program GET L ST. Numbers commonly supplied here are 1.0, 0.1, and 0.01. This scaling factor cannot be used to change the units of measurement as, for example, in changing from feet to metres. When in doubt, use 1/10 of the contour interval as the scaling factor.

15. Request: GIVE LOWEST CONTOUR LINE Z-VALUE WANTED.
Response: Supply the lowest (algebraically smallest) elevation for which a contour line should be shown on the output. If this elevation lies below any point on the generated surface, there will be no output for that contour and, hence, no difficulty will arise. However, some time will be saved if the value supplied here is not lower than the lowest contour that will appear in the contouring region.

16. Request: GIVE HIGHEST CONTOUR LINE Z-VALUE WANTED.
Response: Supply the highest (algebraically largest) elevation which is to be shown. Again, if this number is higher than the highest elevation in the contouring region, no harm is done except that some time is wasted

17. Request: GIVE CONTOUR INTERVAL
Response: Supply the contour interval as a positive number.

18. Request: DO YOU WANT TO SMOOTH THE SURFACE.
Response: If the surface must pass through the data points, type no. The program then assumes SMOOTH=0.0 and jumps to Request 20. If the data points appear too rough, the figure field surface may be smoothed into a relatively flat-lying surface which is not greatly distorted by the sharp bends and gradients evident in the data points. Such a smoothed surface will not pass through the data points. But a

smoothed surface is not necessarily "smooth" looking. Although its gradients are more gentle, it may look just as irregular as a surface produced without smoothing. If smoothing is desired type YES and give the degree of smoothing

19. Request: GIVE DEGREE OF SMOOTHING 0.0-1.0

Response: Supply a number less than 1.0 which indicates the degree of smoothing desires. Care must be taken in that a high degree of smoothing may lead to a flat but discontinuous surface (plateaux). A degree of smoothing greater than 0.7 is not recommended. A degree of smoothing greater than or equal to 1.0 will produce nonsense.

20. Request: IMAGE OF DATA TAPE

(Name of data point file)
(Number of data points)
(AP)
(BP)
(CP)
(DP)
(A)
(B)
(C)
(D)
(DEL X)
(DELY)
(SMOOTH)
(FACTOR)
(Lowest contour elevation)
(Highest contour elevation)
(Contour interval)
ARE THESE PARAMETERS OK ?

Response: Type YES if the parameters look correct. Except for DELX and DELY, all names and numbers should be the names and numbers which were just typed (although not necessarily character for character equivalents). If there is an error, type NO and the program will jump to request 1. All Requests will then be repeated. If YES is typed, the paper tape is punched with the above information.

21. Advice: BEGIN NEXT JOB

Response: The program now jumps to Request 1 to begin again. Run out the paper tape, remove it and label it. If there are no more jobs, type C which will cause the program to return to the DEC MONITOR.

3. Preparation of data tape off-line

This same information may be punched up on the off-line teletype. In this case the operator must pay close attention to the tape format. Also, DELX and DELY must be calculated by hand.

First shut off the punch and thin push carriage return. Now turn on the punch and punch a leader with SHIFT, CONTROL, REPEAT and P depressed simultaneously. Now type a space followed by the 9 characters of the data point file name. Type carriage return twice (skipping one line) and type the number of data points in this file at an integer (no decimal point) right justified in a ten character field width. Type two carriage return (to skip a line). Type the values of AP, BP, CP and DP on successive lines. These and all succeeding parameters may be anywhere in a ten character field width and they must contain a decimal point. Again, type two carriage returns (to skip a line). Type the values of A, B, C and D on successive lines. Again, type two carriage returns (skip a line). Type the values of DELX and DELY, the grid cell dimensions, on successive lines. Type two carriage returns (to skip a line) and type the degree of smoothing and the Z scaling factor on successive lines. Type two carriage returns (to skip a line) and type the lowest contour elevation, the highest contour elevation and the contour interval on successive lines. Terminate with a carriage return and run out the tape with SHIFT, CONTROL, REPEAT and P depressed simultaneously.

Example:

Ten character field width

I GS2 SRC	Name of data point file
286	Number of data points
1ØØ.ØØØ	AP
415.ØØØ	BP
1ØØ.ØØØ	CP
415.ØØØ	DP
1ØØ.ØØØ	A
415.ØØØ	B
1ØØ.ØØØ	C
415.ØØØ	D
6.3ØØØØØ	DELX
6.3ØØØØØ	DELY
Ø.ØØØØØØ	SMOOTH
Ø.1ØØØØØ	FACTOR
-6.5ØØ	Lowest contour elevation
3.ØØØ	Highest contour elevation
Ø.5ØØ	Contour interval

Operational Instructions - GETLST

Program name	GETLST
Function	To read point or line data from internal E.C.U. storage and write it in character form to a DECTAPE file.
Written by	D.S. Connelly

1. Introduction

For editing efficiency as well as for ease of input, the data point (X, Y, Z) lists to be read in by the figure field generating routines must be in character form (3F12.5). Most data point lists will be compiled by, and derived from the standard internal E.C.U. system. GETLST will translate point data and line data from the standard internal E.C.U. format into a named file of (3F12.5) X, Y, Z co-ordinate triplets in character form. In converting line data to X, Y, Z co-ordinate triplets, the program will retain only each Nth point along each line, where N is a number selected by the operator. This crude filtering allows the user to reduce the large number of data points on a contour line to a reasonable number of points for figure field generation. A means is also provided for merely counting the number of points to be generated without writing an output file.

2. Program setup

Mount the input MAGTAPE on unit 5 with the ring removed. Mount a file-oriented DECTAPE with sufficient free space on unit 1. Mount the figure field operations DECTAPE on unit 2. Initiate the DEC MONITOR and type E GETLST.

Q1. ENTER INPUT SHEET NUMBER
R1. 1.0

Give the number of the sheet (file) containing the data points. If the sheet cannot be found, the question will be repeated.

Q2. GIVE NAME OF OUTPUT FILE (6 LETTERS + SRC)
R2. CHALK4SRC

Type the file name as 6 letters and/or numbers and/or spaces followed immediately by the letters SRC. If no name is given, the program assumes that there is to be no output and it processes the input file merely to determine how many points

Q5.3. RETAIN EVERY N-TH POINT ON A LINE
WHAT IS N

R5.3. 50.0
The first point of every curve is listed. In addition every Nth point along that line is retained. Any positive integer (with a decimal point) may be typed.

The corner point locations will be typed out. They are not stored on the output file.

When the end of sheet is reached, the output file will be closed and the number of points that it contains will be typed.

The program then restarts with Q1. Carry on if more files are to be processed. If nothing remains to be done, type ↑C to return to the DEC MONITOR.

Operational Instructions - GRATGN

Program name	GRATGN (MK 1A)
Function	To generate a grid or a graticule in any one of the current available projections
Written by	D. Vicars

1. Introduction

This program uses a general subroutine which generates a grid in minimum intervals of 1km or a graticule in minimum intervals of 1 degree.

2. Program setup

The program uses one magnetic tape for output; this must be mounted on unit 6, write enabled. The program then asks the questions necessary to name the output file and define the grid or graticule.

Q1. IS THIS THE FIRST FILE ON O/P TAPE
R1. NO

This causes the output tape to be positioned just past the last data on the output tape.

If the reply is 'YES', then the output tape is positioned at the beginning and question 2 asked.

Q2. ENTER TAPE NUMBER
R2. MT99.

The program now asks for the permanent file details:

Q3. ENTER PERMANENT FILE DATA
PROJECT REF
R3. N102

This is a 5 character reference for that project.

Q4. DESCRIPTION
R4. GULF OF ADEN

The first 25 characters are retained.

Q5. JOB NUMBER
R5. 73.

This is intended to be a numeric project reference for costing.

Q6. SHEET REFERENCE
R6. 2.

This is a 10 character reference code for the sheet; this could be a complete O.S. reference, e.g. SU 424336, or a UTM reference.

Q7. ENTER DATE AS DD-MM-YY
R7. 11-Ø3-7Ø

The reply is given some simple checks; if any of them fail the message 'ERROR' is typed and the question repeated.

Q8. ENTER YOUR INITIALS
R8. CW

Q9. ENTER RUN DESCRIPTION
R9. GRATICULE AT 1 DEGREE INTERVAL

The program then types the message 'OUTPUT TO FILE n', n being the file number.

Q10. WHICH PROJECTION
R10. MERC

The reply must be one of the following, otherwise the message 'NAME UNKNOWN' is typed and the question repeated.

Accepted projections:

Gall	GALL
Modified Gall	MGALL
Mercator	MERC
Equidistant cylindrical	EDCYL
Equal area cylindrical	EACYL
Equidistant zenithal	EDZEN
Equal area zenithal	EAZEN
Orthomorphic zenithal	OZEN
Orthomorphic conical	OCON
Kilometre grid	GRID

Q11. ENTER O/P SCALE
R11. 4ØØØØØØ.

Q12. ENTER GRATICULE INTERVAL IN DEGREES
R12. 1.

Alternatively, if the reply to Q10 was 'GRID', Q12 becomes:

Q12a. ENTER GRID INTERVAL IN KMS.

R12a. 1Ø.

Q13. ENTER CORNER COORDS
SW CORNER

R13. N8.
E42.

This is repeated for the NW, NE and SE corners. If a graticule is being generated the coordinates must be in degrees preceeded by 'N' or 'S' and 'E' or 'W' and in the range Ø. to 18Ø., otherwise they must be in kilometres.

Operational Instructions - IBMIN

Program name	IBMIN
Function	To convert an IBM tape from standard Fortran IV format, as defined by IBM for the 1401, 7094, etc., (and also used by the Atlas, CDC6600, DKF9) to the format required by the D.E.C. Fortran system.
Written by	T. Lang

1. Introduction

Many computer operating systems will read and write magnetic tape using BCD recording (i.e. Binary Coded Decimal characters) according to the original IBM specification. However, the D.E.C. operating system uses a unique format not compatible with other systems. The IBMIN program will process a tape in the original IBM format and do the necessary conversions to provide a tape which can be read directly by our own programming system.

2. Program setup

The program will start with the message

IBM-IN PROGRAM IA
MOUNT BCD I.P. TAPE ON MT5, O.P. TAPE ON MT6
PAUSE

The input tape should be write-disabled, the output tape write-enabled. When the tapes are loaded, type ↑P to continue. The program will then ask a series of questions:-

Q1. ENTER INPUT TAPE SPECS
- DENSITY (0./1./2. = 200/556/800 BPI)
R1. 0.

Q2. - PARITY (0./1. = EVEN/ODD)
R2. 0.

The replies to questions Q1, Q2 will depend upon the computer which produced the tapes.

Q3. - CHARS PER RECORD (0. IF VARIABLE)
R3. 120.

If the number of characters per record is variable, or not known, a reply of zero should be made. If the number of characters is specified, the program will check to ensure that every record has exactly this number.

The D.E.C. system imposes a maximum theoretical limit of 1268 characters per record, and IBM-IN will correctly process records up to this size. However, there is an error in the D.E.C. system which prevents the reading of records of more than 133 characters. If this lower limit proves an embarrassment (it is still more than the standard length of 120 characters employed by most line-printers) it could, with some effort, be overcome.

Q4. -NO. RECORDS (0. TO END ON EOF/EOT)
R4. 0.

If the number of records is not known, a reply of zero should be made. The program will then continue processing until an End of File mark or an End of Tape mark is sensed on the input tape.

Q5. DO YOU WISH ALL INPUT ERRORS TO BE IGNORED
R5. NO

If the reply is NO the input tape will be checked, and a parity error or wrong number of characters will give rise to an error message. A reply of YES may be used to recover as much useful information as possible from a tape known to be faulty.

Q6. DO YOU WISH TO SPECIFY A NON-STANDARD CHARACTER SET
R6. NO

The original IBM specification defines which binary numbers are to represent which printable characters. If the input tape adheres to this specification, give a reply NO, and the program will then start processing the input tape. If the reply is YES, the user will be asked to specify his own character set. This will be indicated by the message

LOAD DATA TAPE IN PTR
PAUSE

The paper tape defining the new character set (punched according to the convention described in section 4) should then be loaded in the paper tape reader. When the user types ↑P the paper tape will be read.

The program will then commence processing by rewinding both tapes. It will then read one record at a time from the input tape, apply the necessary transformations, and write the transformed version to the output tape. If the user has specified the exact number of records on the input tape, the end of processing will be signalled by the message

RECORD 1076 REACHED

Otherwise processing will end with either

EOF ON INPUT TAPE AFTER 2944 RECORDS

or

END OF INPUT TAPE AFTER 3498 RECORDS

In all three cases an EOF marker will be added to the end of the output tape. The program will finally rewind both tapes, and terminate with the message

STOP

3. Error messages

If either of the tape units is not switched on-line; or the output tape is not write-enabled, this will give rise to the message

IOPS4

Switch the faulty unit to the correct condition, and type ↑R to continue.

Unless the user has specifically asked for errors on the input tape to be ignored (Q5), each record will be checked. There are three possible error messages

INPUT PARITY FAIL ON RECORD 97
PAUSE

The program will attempt to read the record 10 times before signalling the error. Type ↑C to abandon the program, ↑P to continue. If the program is continued the faulty record will be ignored.

INPUT RECORD 637 HAS 117 CHARS
PAUSE

This error will only be signalled where the exact number of characters per record has been specified. The faulty record will be ignored if the program is continued.

INPUT RECORD 1742 HAS ILLEGAL CHAR *16
PAUSE

The program will check every 6-bit character on the input tape, before converting it to the internal D.E.C. format. The "*" character indicates that the value of the illegal character is given in octal notation. If the program is continued, the faulty character is replaced by a <space> character.

The program also checks the output tape as it is produced. There are two possible error messages -

OUTPUT PARITY FAIL ON RECORD 239
STOP

or

END OF OUTPUT TAPE REACHED DURING RECORD 2373
STOP

In both cases the program will be abandoned.

The program will also perform certain checks on the hardware. There are two possible error messages -

CATASTROPHIC FAIL NO. 3 IN RECORD NO. 1011
STOP

and

INPUT RECORD TOO LONG
PAUSE

In the latter case the program may be continued, and the faulty record will be ignored. In both cases the assistance of a programmer should be sought.

4. Non-standard character set specification

The user is given the option (Q6) of specifying a non-standard IBM character set. The new specification must be supplied as table of 64 entries, one entry for each of the 6-bit numbers *00 to *77 (octal). Each entry consists of a 3 digit octal number, defining the equivalent D.E.C. character in 7-bit ASCII (as specified in the PDP9 Advanced Software System hand-book, Appendix A). Illegal 6 bit characters should be given the 7-bit equivalent of 177 (octal).

The new table should be punched on paper tape, one 3 digit entry per line. (The first line may be preceded by a line-feed character, but should not be preceded by a carriage-return character).
e.g.- 177, 061, 062, 063, etc.

Operational Instructions - IBMOUT

Program name IBMOUT

Function To convert an IBM tape produced by the D.E.C. Fortran system to the true IBM tape format accepted by most other computers.

Written by T. Lang

1. Introduction

This program performs exactly the reverse function to the program IBMIN. A 7-track magnetic tape produced on the PDP9 by a Fortran program can not be read on other computers because the DEC operating system employs its own unique character set and also inserts various markers on the tape. The IBMOUT program removes these markers and converts the data to the standard IBM character set. The resultant output tape can thus be transferred to other computers (IBM1401, 7094, etc., Atlas, KDF9, CDC6600, etc.).

2. Program setup

Mount the D.E.C. produced input tape on MT5 (write-disabled), the output tape on MT6 (write-enabled). The program starts by printing the caption

IBMOUT PROGRAM

and then asks a series of questions on the teletype:

Q1. ENTER NO. CHARS PER RECORD (0. = VARIABLE).
R1. 120.

If an exact number of characters per record is specified, the program will check that every record has exactly this number of characters.

Q2. ENTER NO. RECORDS (0. TO END ON EOF/EOT).
R2. O.

Where zero is specified processing will continue until either an end-of-file mark or end-of-tape is encountered.

Q3. DO YOU WISH TO SPECIFY A NON-STANDARD CHARACTER SET
R3. YES

If the reply is NO the program proceeds immediately to Q4. If the reply is YES the program responds with

LOAD DATA TAPE IN TAPE READER

PAUSE

The user should load the data tape, specifying the character set to be employed on the output tape, in the paper tape reader. The format for this paper tape is described in section 4. When ↑ P is struck on the teletype the program will read the paper tape.

Q4. ENTER O.P. TAPE DENSITY (0. = 200, 1. = 556, 2. = 800 BPI)
R4. 1.

Q5. AND PARITY (0. = EVEN, 1. = ODD)
R5. 0.

Only in exceptional circumstances should odd parity be used with BCD tapes. The program will then rewind both tapes and start processing. Various checks on the data are applied; the error messages which may arise are specified in section 3.

Where the user has specified the number of records to be processed the program will cease processing with the message

RECORD NO. 1560 REACHED

Where the program was instructed to terminate on EOT/EOF one of two possible messages will be given

END OF FILE REACHED ON INPUT TAPE AFTER RECORD 2040

or

END OF INPUT TAPE REACHED AFTER RECORD 10517

Finally the program will add an end-of-file mark to the end of the output tape, and terminate with the message

STOP

2.1 Conversion of the first character in a line

Standard Fortran IV stipulates that when an output record is written to a printer, the first character is not printed, but used to control the spacing between lines. In the DEC operating system, this first character in the record is converted to a < line feed > character, regardless of its original value. This character conversion should take place only when the output is to a printer, and not when output is to a non-printing device (e.g. magnetic tape). However, in the DEC operating system the conversion is made regardless of the device type.

In the standard character conversion table employed by IBMOUT a < line feed > character is specified as a character to be ignored. Thus, in effect, the first character of every line is omitted. However, it is possible to have the < line feed > character converted into some 6-bit equivalent by specifying a non-standard IBM character set, as specified in section 4.

Where the exact number of characters per record is specified, the number specified should not include any characters which will be ignored on output - i.e. not the first < line feed > character.

3. Error Messages

The tape units on the PDP9 will only operate in increments of 3 characters at a time (i.e. 3 6-bit characters which make up 1 18-bit word). If the user specifies in response to Q1 a number of characters per record which is not a multiple of 3, a warning will be given

*** WARNING - NOT A MULTIPLE OF 3 ***

PAUSE

Type ↑P on the teletype to continue the program. Records will then be filled out to a multiple of 3 characters by the addition of < space> characters.

The input tape is subjected to a number of checks:

RECORD NO. 103 MORE THAN 512 WORDS

STOP

The DEC operating system is incapable of generating records of more than 512 words. There is no recovery from this error.

PARITY FAIL ON INPUT TAPE RECORD NO. 1019

PAUSE

Type ↑P to continue. The faulty record is processed (<u>not</u> ignored).

ILLEGAL ASCII CHAR *177 IN RECORD NO. 276

PAUSE

The '*' indicates that the value of the illegal character is given in octal notation. Type ↑P to continue. The faulty character will be replaced by a < space > character.

RECORD NO. 2095 HAS 121 CHARS

PAUSE

This error will only arise where the exact number of characters per record has been specified. Type ↑P to continue. A short record will be filled out with <space> characters. A long record will not be truncated, but may be filled out if necessary to an exact multiple of 3 characters with < space > characters. (Where a record is too long only by virtue of extra <space> characters, the record length will be reduced to the expected length and no error message given. In some circumstances < space> characters are added to the end of records by the DEC operating system).

Some checks are also applied to the output tape as it is generated. All errors are irrecoverable:

END OF OUTPUT TAPE REACHED DURING RECORD NO. 7041

STOP

PARITY FAIL ON OUTPUT TAPE DURING RECORD 609

STOP

CATASTROPHIC FAILURE NO. 2 ON RECORD NO. 644

STOP

This latter error message denotes a serious hardware failure. The advice of a programmer should be sought.

4. Non-standard character sets

The user is given the option (Q3) of specifying a non-standard character set. The standard character set normally employed by the program is detailed in Appendix I.

A non-standard character set must be supplied on paper tape. The tape should contain 128 numbers, each on a separate line. Each number of two octal digits, specifies the number on the output tape which replaces the corresponding 7-bit ASCII character. Characters to be ignored should be given the equivalent value -1. (The ASCII

< line feed> character should generally be given the value, as it is inserted by the operating system at the start of every line). Illegal characters, which are to give rise to an error message, should be given the equivalent value -2.

Appendix I

Conversion from 7-bit ASCII to standard 6-bit IBM equivalent.

ASCII 7-BIT CHARACTER (OCTAL)	PRINTING CHAR	IBM EQUIVALENT (OCTAL)	ASCII 7-BIT CHARACTER (OCTAL)	PRINTING CHAR	IBM EQUIVALENT (OCTAL)
000 - 011		-2	101	A	61
012	<line feed>	-1	102	B	62
013 - 037		-2	103	C	63
040	<space>	20	104	D	64
041 - 043		-2	105	E	65
044	$	53	106	F	66
045 - 046		-2	107	G	67
047	'	14	110	H	70
050	(	34	111	I	71
051	)	74	112	J	41
052	*	54	113	K	42
053	+	60	114	L	43
054	,	33	115	M	44
055	-	40	116	N	45
056	.	73	117	O	46
057	/	21	120	P	47
060	0	12	121	Q	50
061	1	01	122	R	51
062	2	02	123	S	22
063	3	03	124	T	23
064	4	04	125	U	24
065	5	05	126	V	25
066	6	06	127	W	26
067	7	07	130	X	27
070	8	10	131	Y	30
071	9	11	132	Z	31
072 - 074		-2	133 - 177		-2
075	=	13			
076 - 100		-2			

N.B. -1 = ignore this character on conversion
-2 = illegal character - replaced by <space>, after output of an error message.

Operational Instructions - LINLAB

Program name	LINLAB
Function	To plot on the light spot projector the feature code values or serial numbers of line data
Written by	T. Lang

1. Introduction

The Gealux provides the facility to expose point symbols on film. In particular these point symbols may be digits. The LINLAB program uses this facility to plot labels in hopefully aesthetically pleasing positions alongside lines. A label may be either the value of a particular feature code of the line, or the serial number of the line as it is stored on magnetic tape.

This program should be regarded as experimental. It was basically just a test-bed for the basic labelling subroutine. Nevertheless it may have useful applications in its raw form.

2. Choice of label positions

The algorithm used to select the best position along the line in which to plant the label is described in detail in Programmer's Note "Line Labelling by Gealux". Firstly the program searches for a part of the line which is straight. The "straightness" of a label position is determined by joining the first and last character positions of the label by a straight line. The maximum displacement of any of the centre characters from this line is then determined, and this displacement used as the measure of straightness. The program user is allowed to specify the maximum preferable displacement. If two label positions are found, both with a displacement less than the acceptable limit, the most horizontal of the two positions is taken.

An exception to this procedure is made in the case of truly straight lines - i.e. a line with two successive data points separated by a distance greater than the length of the label. Such a "straight" position is preferred to all others.

If a line is too short to produce a suitable label position, the label is planted on the straight line joining the end points. If the end points are less than 2mms apart a horizontal label is produced above the start position.

The user may ask to have the label displaced above or below the line with which it is associated. This feature could be used to position several labels close to the same line.

3. Program setup

The input data tape should be mounted on MT5 (write disabled). The Gealux should be set up with the correct point exposure and magnification. The program starts with the heading

LINE LABEL PROGRAM IA

and then goes on to ask

Q.1 ENTER SHEET NO.
R.1 2.

The program will search for the requested sheet, and if it is found print out a summary of the heading details. If the sheet is not found

SHEET 2 NOT FOUND

will be generated and question 1 repeated.

Q2. ENTER SCALE
R2 a) N 0.5 or any of the usual alternatives -
R2 b) D 1.0
R2 c) A 25000.
R2 d) X50.0
Y50.0

Q3. CORNER PT. SYMBOL NO.
R3. 34.

This is the number of the symbol on the Gealux symbol disc. Enter -1. to suppress the corner marks.

Q4. NO. OF LABEL FEATURE CODE (O. FOR SERIAL NO.)
R4. 1.

(i.e. use the values of feature code no. 1 of each line as the labels)

Q5. SYMBOL SPACING (MMS)
R5. 1.5

Q6. DISPLACEMENT OF LABEL FROM LINE (MMS)
R6. 2.0

Specify 0.0 to centre the label on the line, or a negative distance for a label below the line.

Q7. MIN DEVIATION OF LABEL FROM STRAIGHT (MMS)
R7. 2.0

If a large deviation is specified labels will tend to be curved but on a horizontal base, whilst a small deviation will tend to favour straighter labels, but not necessarily on a horizontal base.

Q8. DISTANCE BETWEEN LABEL TRIALS (CMS)
R8. 2.5

To save computing time the labelling routine does not inspect the entire line for a suitable label position, but makes trials at intervals, as selected by the user. It will usually be convenient to select a trial interval sufficiently small that the program can compute label positions just a little faster than the Gealux can plot the labels.

After the response R8 the program will start processing the data, and the "start" button on the Gealux should be pressed.

At the end of the processing, the message will be given

END OF SHEET
PROCESSING TOOK 33 MINS 21 SECS

Q8. ANOTHER IP SHEET FROM THIS TAPE
R8. NO

If the reply is YES the program will loop back to Q1.

Q9. ANY MORE IP TAPES
R9. YES

If the reply is NO the program will STOP, if the reply is YES the program will instruct

LOAD NEW TAPE
PAUSE

When the new input tape is mounted, type ↑P to continue. The program will loop back to Q1.

Operational Instructions - LPP

Program name LPP

Function To plot Gealux line symbols on film

Written by T. Lang

1. Introduction

Before using the Gealux to draw a map it may be necessary to investigate the appearance of various line symbols, or combination of line symbols, to see if they are visually satisfactory. The Line Placement Program facilitates such an investigation. The user may divide a rectangular plotting area into any number of rows. He may then specify any line symbol to draw a line along any row. Any number of symbols may be superimposed on one row.

2. Program setup

Set up Gealux, with its origin near the bottom left corner of a rectangular piece of film (leaving a small margin).

The program starts by printing its title, and then asking a series of questions on the teletype:

Q1. ENTER LINE LENGTH IN CMS
R1. 20.0

Q2. ENTER HEIGHT OF FILM IN CMS
R2. 10.0

Q3. ENTER TOTAL NUMBER OF LINES
R3. 2.

Q4. ENTER SYMBOL NO. FOR REFERENCE MARKS
R4. 30.

These questions establish the layout of the line positions on the film. The corner reference marks are put down as point symbols. The above example would result in layout

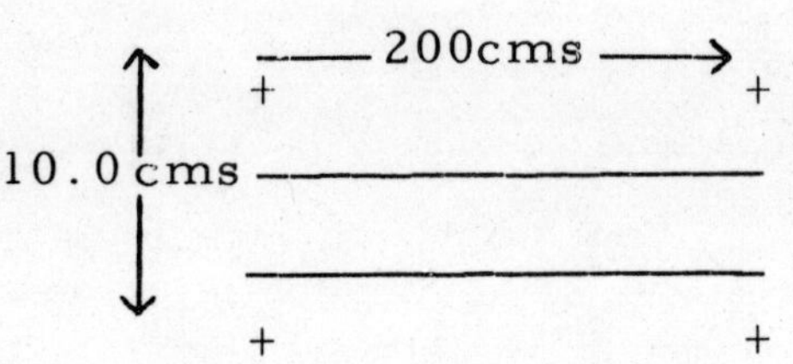

The program goes on to ask for the various line specifications

Q5. ENTER LINE ROW NO.
R5. 2.

Rows are numbered from top to bottom, starting with row 1

Q6. ENTER LINE SYMBOL NO.
R6. 6.

This is the position of the symbol on the symbol disc

;
Q7. ENTER LINE PECK NO.
R7. O.

This sets up the pecked line generator. Specify O. for a continuous line, 1. to 9. for pecked lines.

The program draws the line as specified, and then repeats from question 5. Lines may be superimposed by specifying different symbol (and peck) numbers with the same row number. The user may change the symbol magnification and peck pattern size between lines.

To terminate the program, type ↑C at any time. To start another test with new film, specify a row number of O. (Q5.); the program will restart from Q1.

Operational Instructions - LSTCOD (MKIA)

Program name	LSTCOD
Function	To list on paper tape the feature codes of each line or point item on a tape in standard internal ECU format.
Written by	D. Cranch

1. Introduction

The program is intended to provide a facility for checking the feature codes of all items in selected files at a minimum cost of machine time by punching the feature codes on to paper tape for subsequent checking.

To assist in the identification of each item the start and end x- and y- co-ordinate of each line item are punched, and the x- and y-co-ordinates of each individual point item or corner point item are punched.

The data for each item on the paper tape are preceded by a serial number which is initialised to 1 and incremented by 1 for each line or point item punched. This serial number provides an indication of erroneous data on the input tape (see below, paragraph 3).

2. Program setup

The input tape must be mounted on MT5, write disabled. The program begins by typing its name, mark number, and date of issue, followed by a a reminder to mount the input tape on MT5. This is followed by the questions:

Q1. ENTER TAPE NUMBER (UP TO 5 CHARACTERS)

R1. MT123
This tape number appears on the paper tape as:
TAPE NUMBER MT123
and provides a reference on the eventual p/o.

Q2. ENTER FILE NUMBER S. MAX NUMBER OF ENTRIES IS 50
END WITH 99999.

R2. 10.
12.
-16.
110.
99999.

The numbers of the files for which a listing is required are entered here, preferably in ascending order. The program understands a negative entry as meaning all files from the preceding number to the negative number. Thus, in the example, a listing will be punched for file numbers 10, 12, 13, 14, 15, 16 and 110. All numbers must be terminated with a dot, and the list is terminated by entering 99999.

The maximum number of entries which can be made is 50; the above example counts as four entries. After entering 99999. the program searches the list of file numbers for duplicates. If a duplicate is found, e.g. suppose in the above example that 110. had been entered as 10., the following message and command are typed:

FILE NUMBER 10 IS DUPLICATED

Q3. RE-ENTER FILE NUMBER BEFORE FIRST FAULTY ONE

R3. -16.

Q4. CONTINUE ENTERING FILE NUMBERS

R4. 11Ø.
99999.

Q5. CORRECTED LIST IS
10
12
-16
110
IS LIST O.K.

R5a YES. The program continues as described below.
R5b NO. The program loops to Q3.

When the list has been cleared of duplicates, the program attempts to find each entered file after punching, e.g.

FILE NUMBER 10

If unsuccessful the program types and punches

FILE NUMBER 91 NOT FOUND
and attempts to find the next required file.
If the file is found the program punches the following data:

for a line item:

1st line: Serial number, start x- and y- co-ordinates, end

x- and y- co-ordinates, number of feature codes, feature codes (up to four codes).

2nd and subsequent lines: (if necessary) ten further feature codes per line.

for a point item:

1st line: serial number, x- and y- co-ordinates, number of feature codes, feature codes (up to six codes)

2nd and subsequent lines: (if necessary) up to ten further feature codes per line. When the end-of-file marker is found, the program punches:

END OF FILE NUMBER 91

and then attempts to find the next requested file. When all the requested files have been dealt with the program asks:

Q6. ANY MORE FILES ON THIS TAPE

R6. (a) YES, program goes to Q2, or (b) NO, program punches
END OF TAPE NUMBER MT123
and goes to Q7.

Q7. ANY MORE INPUT TAPES

R7. (a) YES, program goes to Q1, or
(b) NO, program types and punches:
FINISHED.

3. Indications by serial number on paper tape of faulty data on input magnetic tape

Normally the serial number will commence with a value of 1 and increase in value by 1 for each item punched.

If the value of the serial number increases by 2 (or more) between successive items, this is an indication that 1 (or more) lines have been encountered not having an end point. The data actually punched should be correct in all respects save omission of the incomplete line.

If 2 (or more) successive values of the serial number are identical this is an indication that 1 (or more) lines have been encountered not having a starting point. In this case the start x- and y- co-ordinates of the second (and later) items having the same serial number should be ignored.

Operational Instructions - MERGE

Program name MERGE (MK 1A)

Function To merge feature codes punched offline on paper tape with features from magnetic tape.

Written by D. Vicars

1. Introduction

The codes on paper tape may be punched in completely free format, with as many numbers to a line as desired. Errors in the tape can be erased during punching by the normal use of the teletype "RUBOUT" key, or after punching with DEC Editor program.

2. Free format input

In this system integer numbers can be terminated by any non numeric character, and can be separated by any number of spaces. Thus the user can either compress as many numbers as possible on a line, or alternatively arrange them neatly in columns. Lines may contain comments to help the user to identify sections of data, but once alphabetic data is encountered no further codes can be entered on that line.

```
e.g. 1312   768    1   14  Section 1.
       18  40    121       Section 2.
```

would be read as:

```
1312
 768
   1
  14
  18
  40
 121
```

Numbers are normally assumed to be a maximum 5 digits, so that if several numbers are run together without any intervening spaces the program will split them up in 5 digit groups. The program will reset this 5 digit counter if it encounters an equals sign followed by a numeric character in the range 1 to 6 in the input data.

```
e.g.  26128 = 2-157459470.  Section 3.
      =573942.          End of Section 3.
```

would be read as:

```
26138
  -15  (max no. of digits now 2)
   74
   59
   47
    ø
73942  (max no. of digits now 5)
```

If a number is terminated by the character "/" then it is ignored (i.e. erased) and the text one read even if it is on the next line.

```
e.g. 3151 / 272ø  273ø
would be read as
    272ø
    273ø
```

("/" can not be used to erase a change in the digit counter)

The program will also read numbers as octal (i.e. numbers to the base 8 instead of 10) if they are preceeded by an exclamation mark ("!"). A maximum of 6 digits are considered. This cannot be reset. The number can be terminated by any non octal digit (i.e. 8 or 9 as well as non numeric characters)

```
e.g. = 3  125734!  2567! 1øø! 2øø / !177
would be read as:

     125  (3 digits only considered)
     734
    1399  (Decimal representation of! 2567)
      64  (    "              "        "! 1øø )
     127  (Decimal! 177 ! 2øø erased)
```

3. Program setup

Mount the input tape on MT5, (write disabled), the output tape on MT6, (write enabled), and the paper tape in the reader. The program will then type its name, mark number and date of issue. It will then ask a series of questions.

Q1. IS THIS THE FIRST FILE ON O/P TAPE
R1. NO

This will cause the merged file to be output at this end of the data already on the output tape. If the reply is "YES" then the program will ask question 2 and commence writing at the beginning of the output tape.

Q2. ENTER TAPE NUMBER
R2. MT 8ø.

The reply must be "MT" or "DT" followed by the number terminated by a decimal point.

Q3. ENTER INPUT FILE NO.
R3. 3.

The program will search for this file on the input tape. If it cannot find it the message "FILE 3 NOT FOUND" is typed and Q3 repeated.

Q4. ENTER DATE AS DD-MM-YT
R4. Ø9-Ø2-7Ø

If the program cannot understand the reply it will type "ERROR" and repeat the question.

Q5. ENTER YOUR INITIALS
R5. ASK

Q6. ENTER RUN DESCRIPTION
R6. SECOND MERGE CODES 2 TO 5

Q7. LOAD TAPE IN READER
PAUSE ØØØØØØ
R7. ↑P

The paper tape must be loaded carefully in the reader and the white button above it momentarily pressed before typing "CTRL" and "P" together.

Q8. MERGE FROM CODE NO.
R8. 2.

Q9. TO CODE NO.
R9. 5.

These replies would cause 4 numbers to be read from the input stream for each feature, they would overlay codes 2, 3, 4 and 5 leaving the others untouched; if the feature had less than 5 codes before merging it will have 5 after. The two numbers must be terminated with a decimal point, R8 must be less than or equal to R9 and both less than 51. If not then an error message is given and the questions repeated

Q10. FOR HOW MANY FEATURES
R10. 2ØØ

The scheme laid out in Q8, Q9 will be applied to the next 2ØØ features on the input file. If the end of the input file has not been reached Q8 etc. will be repeated. When the end of the input file is reached, the program will type out the next line of characters from the paper tape. If the user has inserted a suitable line of comments at the end of his data, the typing of these comments will confirm that the correct number of codes was read from the paper tape. This is followed by counts of the number of features and codes read, and the time taken for the run.

Q11. DO YOU WANT TO MERGE ANOTHER FILE
R11. NO

This will cause the output tape to be closed, the input tape to be rewound, and the run terminated. If the reply is "YES" the program will re cycle from Q2 for another file.

Errors

OTS 11 The input paper tape has ended before all the features on the input tape have been processed. No operator action is required; the error is irrecoverable.

IOPS4 Device (i.e. tape unit or paper tape reader) unready. Make device ready and type "CTRL" and "R" together.

Operational Instructions - MTCPY

Program name MTCPY

Function To copy and verify IBM tapes

Written by T. Lang

1. Introduction

Where the data stored on a magnetic tape is the result of much human or computer effort it is sensible to guard against mishaps by making a duplicate copy. MTCPY will deal with tapes recorded in any of the usual densities and of odd or even parity. It is also possible to produce the copy in a different density and parity from the original.

The input tape is checked for parity errors, and correct number of characters in each record. The facility is also provided to read, and hence check, the input tape, without writing an output tape.

Once an output tape has been produced it can be verified. This is accomplished by reading through the new tape record by record, and comparing it against the original.

2. Program setup

The program starts by printing a heading, and listing the numbers which must be employed by the user to indicate tape parity and densities -

MT COPY PROGRAM
DENSITY O. = 200 1. = 556 2. = 800 BPI
PARITY O. = EVEN 1. = ODD

This is followed by the question

Q1. DO YOU WANT TO COPY STANDARD TAPES
R1. YES

The reply YES indicates that the user wishes to copy a tape produced in the normal internal ECU format. In this case the program knows what parity and density are required and questions Q2 to Q10 are all omitted. It is assumed that the input tape is loaded on MT5 (write disabled) and the output tape on MT6 (write enabled). The program immediately enters the copying process.

A reply of NO indicates that a non-standard tape is to be copied. A series of questions are asked to elicit the details -

Q2. ENTER INPUT TAPE SPECS.
- TAPE UNIT
R2. 3.

This tape unit should be write disabled.

Q3. - DENSITY
R3. 1.
Q4. - PARITY
R4. 1.
Q5. - CHARS PER RECORD (0. IF VARIABLE)
R5. 120.

If there are an identical number of characters in every record, the program will check for this on the input tape. If the number of characters per record is not known, or is variable, a reply of zero should be given. (The maximum size of record is 7500 characters).

Q6. NO. RECORDS (0. TO TERMINATE ON EOF/EOT)
R6. 1077.

If the number of records is not known, zero should be specified, and the copy process will be terminated when an EOF (end of file mark) or EOT (end of tape mark) is encountered.

Q7. DO YOU WISH TO IGNORE ALL INPUT ERRORS
R7. NO

If the reply is NO the program will check for errors on the input tape, and if any are found print warning messages. If the reply is YES the program will ignore these errors and do its best to produce an output tape. This option may be used to recover as much useful information as possible from a faulty tape.

Q8. ENTER OUTPUT TAPE SPECS - TAPE UNIT (8. IF NONE)
R8. 7.

The output tape unit should be write enabled. If a unit number of 8 is specified no output tape will be produced, but the input tape merely read through and checked for parity errors or wrong number of characters in a record. If the facility is to be used, questions Q9 and Q10 will be omitted.

Q9. - DENSITY
R9. 1.
Q10. - PARITY
R10. 1.

The program will then rewind both input and output tapes, and start the copy process. If any errors are detected during the copy process warning messages (described in section 3) will be printed.

One of three messages will indicate that the copy process is complete. If the exact number of records has been specified (Q6), the message will be
RECORD NO. 1077 REACHED ON INPUT TAPE
Otherwise the program will read until EOF or EOT are encountered, giving rise to
INPUT TAPE TERMINATED BY EOF AFTER 2073 RECORDS or
INPUT TAPE TERMINATED BY EOT AFTER 1397 RECORDS

If the input tape is terminated by an EOF, an EOF will be written to the output tape also.

If there was no output tape, or if input errors were ignored, the program will terminate with the message.
STOP
Otherwise the question will be asked

Q11. DO YOU WISH TO VERIFY THE O.P. TAPE
R11. YES

If the reply is NO the program will terminate. If the reply is YES the program will rewind both tapes, and read through them again, comparing each pair of records. Any errors will give rise to warning messages, as described in Section 3.

As with the copy process, one of three messages will indicate that the verification process is complete -
VERIFICATION FINISHED AFTER RECORD NO. 1077
or
INPUT TAPE TERMINATED BY EOF AFTER 2073 RECORDS
OUTPUT TAPE ALSO HAS EOF
or
INPUT TAPE TERMINATED BY EOT AFTER 1397 RECORDS
In all three cases the program will terminate with the message
STOP.

3. Error Messages

3.1 Errors during copy phase

Two messages may be typed to indicate errors on the input tape (provided the option to ignore these errors is not being utilised).
INPUT RECORD NO. 324 HAS 117 CHARACTERS
PAUSE
This will appear only where the number of characters per record has been specified in reply to Q5. Type ↑C to abandon the program, ↑P to continue. If the program is continued the faulty record will not be copied to the output tape.
INPUT TAPE PARITY FAIL/BAD TAPE ON RECORD NO. 327

PAUSE
The program will make 10 attempts to read the faulty record before giving the error message.

The hardware will also check the output tape as it is being written. An error will give rise to the messages.
OUTPUT TAPE PARITY FAIL/BAD TAPE ON RECORD NO. 76
STOP
or
END OF OUTPUT TAPE REACHED DURING RECORD NO. 2074
STOP

In both cases the program will be forcibly abandoned.

3.2 Errors during verify phase

A parity failure on either tape will be indicated by
INPUT TAPE PARITY FAIL/BAD TAPE ON RECORD NO. 1073
PAUSE
or
PARITY FAIL/BAD TAPE ON OUTPUT TAPE RECORD NO. 2076
PAUSE

If corresponding records from the two tapes are found to contain different numbers of characters this will give rise to
RECORD NO. 1799 INPUT 120 CHARS, OUTPUT 117 CHARS
PAUSE
If the records do not contain identical information -
RECORD NO. 993 HAS COMPARISON ERROR
PAUSE
If an EOF is detected unexpectedly on the output tape -
EOF ON O.P. TAPE AFTER RECORD NO. 434
If the input tape was terminated by an EOF, the program checks that the output tape is also terminated by an EOF. If the EOF is missing on the output tape, the verification phase will end with the message
INPUT TAPE TERMINATED BY EOF AFTER 2073 RECORDS
BUT O.P. TAPE DOES NOT HAVE VALID EOF

3.3 Miscellaneous errors

If either of the tape units are not switched "on-line" or the output tape is not write-enabled, a message will be given.
IOPS4

Switch the faulty unit to its correct condition and type ↑R to continue.

The DEC hardware can only read from or write to magnetic tape in units of 3 characters. If the user specifies a number of characters

per record (Q5) which is not an exact multiple of 3, a warning will be given
*** WARNING *** NOT A MULTIPLE OF 3
PAUSE

The user should type ↑C to abandon the program, ↑ P to continue. If the program is continued the number of characters per record will be rounded up to an exact multiple of 3 (i.e. one or two characters will be added to the end of the output record).

Whilst the program is running it makes a number of checks to ensure that the hardware is operating correctly. If a hardware error is encountered a message of the form
STOP X
will be printed, where X is a non-zero number. The program will be forcibly abandoned. The help of a programmer (preferably the author) should be sought.

Operational Instructions - MTEDIT (MK1B)

Program name MTEDIT

Function To provide rudimentary editing facilities.

Written by D. Vicars

1. Introduction

The program runs in two passes. In the first each feature is read, plotted, if desired, and its serial number and codes typed. The operator then directs the program to take suitable action on it. Available commands are:

(i) Accept the feature
(ii) Delete the feature
(iii) Change (or add) feature codes
(iv) Undo loop. This can be given for a loop only and must preceed one of the other commands.
(v) Skip; to accept current and all remaining features
(vi) Finish, to delete current and all remaining features

The commands are referenced by the first letter used above in the descriptions. At the end of the first pass the program types out lists of the numbers of features to be deleted, changed (with a list of the new codes) and loops to be undone. The program then requests the file updating information and copies the input tape executing the commands stored from the first pass.

2. Program setup

The input tape must be mounted on MT5 write disabled, the output on MT6 write enabled.

The program begins by typing its name, mark number, and date of issue. This is followed by the questions:

Q1. ENTER FILE NUMBER
R1. 7.

The input tape will be searched for the 7th file. If the tape does not contain seven files the program types "FILE 7 NOT FOUND" and Q1 will be repeated. The file description is typed for identification followed by question 2.

Q2. IS THIS OK
R2. YES

The reply "NO" causes Q1 to be repeated so that another file can be selected.

Q3. DO YOU WANT VISUAL O/P
R3. YES

This will cause the program to draw every feature on the display or the geagraph at a scale and tolerance specified in Q4 and Q5. If the reply is "NO" then drawing is inhibited and Q4 and Q5 skipped.

Q4. ENTER DRAWING SCALE FACTOR
R4. .5

This applies to the drawing only and not to the output tape.

Q5. ENTER STAND-OFF DISTANCE IN CMS
R5. 1Ø.

This will cause the pen to move 10cms away from the operator, after each line so that he can see the whole of the line just drawn

Q6. DO YOU WANT OPERATING INSTRUCTIONS.
R6. NO

If the reply is "yes" the program types out about 14 lines of instructions. (See appendix 1)

The program now reads the input tape, plotting features, if desired. After each feature the program types the message:
"FEATURE n LOOP CODES 1) 23 2) 1 3) 0 4) 1510 etc." or, if the feature has no codes "FEATURE n LOOP CODES NONE."

This is followed by a request for an editing command which must be one of the following:

(i) "A" or just 'return' to accept the feature.

(ii) "D" to delete the feature.

(iii) "C" to change or add codes. This will be followed by the message "ENTER NEW CODES", then for each code two numbers are requested, the code position number and the code itself. This procedure terminates when a code position number of 99999 is entered. The program checks that a code position number is either greater than 0 and not more than 50 or 99999 and if it is neither then it types the message "ERROR IN LAST LINE" and expects a correct number. e.g. FEATURE 53 LINE Codes 1) 0 2) 137 3) 42 4) 2 C

```
ENTER NEW CODES
NO.-
> 2.
CODE -
> 43.
NO.-
> 1. (All numbers must end with decimal point)
ERROR IN LAST LINE
> 1.
CODE -
> 3651.
NO.-
> 99999.
```

The feature now has the codes:
1) 3651 2) 43 3) 42 4) 2.
Extra codes are added by addressing the code position immediately following the last code. The total number of codes is then increased by one. Any number of codes up to the maximum of 50, can be added in this way. Any attempt to address a code number beyond this (e.g. 7. if the item has 4. codes) will be rejected by the program with the message:
"CODE NUMBER TOO LARGE".

e.g. FEATURE 71 LINE CODES NONE.

```
> C
ENTER NEW CODES
NO.-
> 1.
CODE-
> 591.
NO.-
> 4.
CODE NUMBER TOO LARGE
> 2.
CODE -
> 0.
NO.-
> 99999.
```

Feature 71 now has the codes: 1) 591 2) 0.

If a code position is addressed more than once, only the last code will be retained. This is useful for overwriting mistakes.

e.g. FEATURE 72 LINE CODES NONE.

```
> C
ENTER NEW CODES
NO.-
```

```
> 1.
CODE-
> 0.
NO.-
> 2.
CODE-
> 40.
NO.-
> 1.
CODE
> 10     (Code 1 is now 10)
NO.-
> 99999.
```

(iv) "U" to undo a loop. If this command is used on a line or point then it will be rejected with the comment "NOT A LOOP". This command must be used before "A", "C" or "S".
e.g. FEATURE 73 LOOP CODES 1) 19. 2) 49

```
> U
> A
```

(v) "S" to accept current and all remaining features on the tape without plotting them.

(vi) "F" to delete current and all remaining features on the tape without plotting them.

If any other reply is given the program will reject it with the comment "COMMAND UNKNOWN" and expect a correct one.

When every feature in the file has been dealt with the program prints out the total number of features followed by three tables: (a) The numbers of features to be deleted, (b) The numbers of features to be changed, together with the new code numbers and (c) The numbers of loops to be undone. The program then completes the second pass of the input tape, applying all the corrections as elicited from the user during first pass, and producing the updated output tape.

Q7. IS THIS THE FIRST FILE ON O/P TAPE
R7. YES.

If the reply is NO, question 8 is omitted.

Q8. ENTER TAPE NUMBER
R8. MT 71.
Q9. ENTER DATE AS DD-MM-YY
R9. 22-03-70

The reply is given some simple checks and if any of them fail then the message "ERROR" is given and the question repeated.

Q10. ENTER YOUR INITIALS
R10. CW
Q11. ENTER RUN DESCRIPTION
R11. 1ST. EDIT RUN.

Operational Instructions - NEARNS

Program name NEARNS

Function Compute the nearest neighbour statistic.

Written by D.S. Connelly

1. Introduction

The nearest neighbour statistic is a measure of the randomness of the spatial distribution of points. For a more complete description of the statistic and its mathematics see: Clark and Evans, "Distance to Nearest Neighbour as a Measure of Spatial Relationships in Populations," Ecology, Vol 37 (1956), pp 391 - 394.

This program is provided so that the nearest neighbour statistic can be used to judge the reliability of the figure field to be produced from a particular data point distribution. A low statistic indicates a low reliability. A statistic of 1.0 or greater indicates an acceptable reliability. See Morrison, Paper of the 1969 ACSM Convention, pp 187 - 194, for a more detailed discussion of the nearest neighbour statistic as an indicator of figure field reliability.

2. Program setup

The data points will be read from a DECTAPE file in the 3F12.5 format which is also used by the figure-field generating routines. Mount the DECTAPE which contains the correct X, Y, Z data point file on DECTAPE drive 1 (write disabled).

The program should be fetched from the figure field operations tape. With this tape on DECTAPE drive 2 initiate the DEC MONITOR and type E NEARNS.

After an introductory heading the program requests operator action.

1. Request: MOUNT PAPER TAPE
   PAUSE ØØØØØØ
   Response: Load the proper figure field parameter input tape in the paper tape reader and type ↑P.

2. Request: MINX = (A), MAX X = (B), MIN Y = (C), MAXY = (D)
   DO YOU WANT DIFFERENT BORDERS
   Response: Since it is anticipated that this program may be used to calculate the nearest neighbour statistic for a

number of trial contouring frames, a facility is provided here for changing the frame borders without reading in a new data tape. If you do not want to change the contouring frame, type NO. To change the borders, type YES and supply the following co-ordinates:

2.1 Request MIN X
2.2 Request MAX X
2.3 Request MIN Y
2.4 Request MAX Y

Response: After each request type the named co-ordinate. Each number must contain a decimal point. If a mistake is made, type ↑T to restart these requests.

The program now searches for all data points falling within the nominated frame border. For each such point the program finds the nine closest data points, whether inside or outside the frame border. It then tabulates nine sums of distances as it searches, one sum each for the distances to the nine closest points. When these sums are complete for all points within the frame borders, the nine average distances to the nine closest points are computed. The ratios of the average distances to the expected distances are also computed. These are the nearest neighbour statistics. When all the work is complete, the results are listed in tabular form. Along with the nearest neighbour statistic, the following quantities are listed: the actual average distances, the expected average distances, the shortest distance to a neighbour of each rank and the longest distance to a neighbour of each rank.

Typing ↑T at any time after data input causes the program to request a new figure field parameter tape. On normal completion of the calculations the program jumps to Request 2 and assumes all the old parameters are still valid. To fetch new a parameter paper tape type ↑T.

Operational Instructions - NGEAG

Program name	NGEAG
Function	To draw on the Geagraph maps stored on IBM tapes in standard internal ECU format.
Written by	T. Lang

1. Introduction

Corner points will appear as crosses on the scribed or drawn output sheet. A feature of this program is the windowing facility described later.

2. Program setup

Before program execution is initiated, the IBM data tape should be loaded on MT5 (write disabled). The Geagraph should be set up ready to draw or scribe.

The program will ask a number of questions, in order to ascertain what is to be drawn:-

Q1. ENTER SHEET NO.
R1. 4.0 (or alternatively just 4.)

Sheet number means the relative number on the tape (starting from 1.) of the sheet which is to be drawn. The reply must include a decimal point. If the reply is 0. then it is assumed that all sheets are to be drawn, and the program loops directly to question Q10.

Q2. SAME SPEC AS LAST TIME
R2. NO

This question is only asked if the previous sheet was juxtaposed (see Q3). If the reply is YES, then the new sheet is drawn to the same specifications as the previous sheet, and all subsequent questions are omitted.

Q3. JUXTAPOSE SHEETS
R3. NO.

If the reply is NO the sheet is drawn with its origin (usually the SW corner) at the zero point on the Geagraph table. If the reply is YES, then the sheet is displaced to its correct position relative to the origin for the whole series of sheets (as specified during JOIN). In this way a number of adjoining sheets may be drawn with their corresponding sides abutting in the correct relative positions.

Q4. DO YOU WANT WINDOW FACILITY
See section 7 for details.
R4. YES

If the reply is NO, the following queries are omitted

Q5. ENTER BOUNDING RECTANGLE AS FRACTION OF SINGLE SHEET SIZE MINIMUM X
R5. 0.5

Q6. MAXIMUM X
R6. 2.5

Q7. MINIMUM Y
R7. 0.5

Q8. MAXIMUM Y
R8. 2.5

Q9. SHIFT WINDOW TO ORIGIN
Q9. NO

Q10. ENTER SCALE

The scale of drawing may be specified in 3 forms, each prefixed by a letter:-

R10(a) N2.0

N is used to indicate that the scale is specified relative to the nominal scale of the sheet. Thus if the sheet was stored on the tape at a scale of 1/1,000,000 (as specified during JOIN), then the above reply would result in a map at 2.0 x 1/1,000,000, i.e. at 1/500,000.

R10(b) D0.5

D is used to indicate that the scale is specified relative to the size of the actual document from which the sheet was digitised. Thus the reply R10(b) would result in a drawing 1/2 of the size of the original.

R10(c) X10.6
Y17.413

The third alternative is to specify the exact dimensions of the sheet along the X.Y axes. The above example would result in a drawing 10.6 cms x 17.413 cms. (In the event of a non-rectangular sheet, the size is adjusted so that the average dimensions, measured over the pairs of corner points, are equal to the nominated sizes).

Q11. ENTER TOLERANCE IN MMS
R11. 0.2

If a point on a line lies closer to the point just drawn than the nominated tolerance it will be omitted (except when it is the end point of the line). Lines of total length less than this tolerance will be omitted entirely. A tolerance of 0.2 or 0.3mms has been found to be satisfactory in achieving good generalisation at reduced scales.

Questions Q12 to Q17 are asked by the standard subroutine called "SELECT" which is used to obtain from the user a specification of which items within the sheet are to be drawn and which omitted.

Q12. ENTER CODE SELECTION
END WITH 99999.

R12. 32.
79.
-164.
99999.

As a reply to Q12 the user must specify a list of feature codes. Each code must be terminated with a decimal point, and the whole list terminated with the code 99999. Items with these codes will be selected for playback.

If the list of codes is empty (i.e. only 99999. has been entered) it is assumed that all items in the sheet are to be drawn. Questions Q13 to Q17 are then omitted.

If the list of codes is terminated not by 99999., but by 99998., then it is assumed that the list so far supplied refers to items to be drawn with a specific radiograph pen, which will be termed PEN 1. The program continues:-

Q13. CODES FOR PEN 2.
R13. 16.
17.
99998.

The user may continue in this way to specify as many different pens as may be required.

Q14. SEARCH FROM FEATURE CODE NO.
R14. 3.

Q15. TO FEATURE CODE NO.
R15. 5.

If each item on the tape had 8 feature codes, only codes 3 to 5 of each item would be compared against the list of selected codes.

Q16. IS SELECTION TO BE INCLUSIVE
R16. YES

If the reply is YES then items with the nominated codes are included for drawing. If the reply is NO then the list of codes is taken to be exclusive i.e. items with these codes are omitted, and all the rest drawn. (Q16 is omitted if more than 1 pen has been requested, since the list of codes must then obviously be inclusive.)

Q17. DO YOU WANT INTERNAL LINES SUPPRESSED
R17. YES

If the reply is YES, then items with more than one code appearing in the list of selected codes will be omitted - e.g. if codes for chalk and clay boundaries had been specified, boundaries between chalk and clay would be omitted.

3. Program Running

Once the reply to question Q17 has been input, no further messages are given, and drawing starts immediately. On the first pass through the program it will be necessary to push the "START" button on the Geagraph.

The program assumes that when drawing starts pen 1 is loaded. If a change of pen is required the program will type

CHANGE TO PEN 2
PAUSE

The user should change to the specified pen, then type ↑P on the teletype to continue.

4. Program Termination

When drawing is complete the program will type a message
DRAWING TOOK 25 MINS 37 SECS

This will be followed by the questions

Q18. ANY MORE SHEETS FROM THIS TAPE
R18. NO

If the reply is YES, the program will loop back to Q1.

Q19. ANY MORE TAPES
R19. NO

If the reply is YES the program will rewind the current IBM tape, and type LOAD NEW TAPE.

PAUSE

To continue once the new type is loaded type ↑P on the teletype. The program will loop back to Q1.

5. Error Messages

If the input IBM tape is not on unit 5 or the unit is not switched "on-line",

IOPS4

will be typed. After resetting the unit correctly, type ↑R to continue.

If in response to Q1 the user specified a sheet number greater than the total number of sheets on the tape, the message
SHEET NOT FOUND
will be output, and question Q1 repeated.

If the program detects a parity error on the input tape it will loop continually, alternately backspacing and attempting to re-read the faulty record.

The reply to Q10 specifying the scale should always be prefixed by N, D or X (followed by Y on the next line). If this condition is not satisfied a message will be typed

PREFIX SCALE BY
N FOR SCALE RELATIVE TO NOMINAL SCALE
D FOR SCALE RELATIVE TO SIZE AS DIGITISED
X THEN Y FOR ABSOLUTE SIZE IN CMS

A number of checks are applied to the list of selected feature codes as they are entered. Three possible error messages may be generated:-

CODE MUST BE LESS THAN 99999.
END ALL NUMBERS WITH DOT (.)
CODE ALREADY IN LIST

After one of these messages the user should re-enter the correct code. Alternatively the selection process may be restarted from the very beginning by typing ↑T on the teletype.

A maximum of 50 codes may be entered, associated with up to 50 different pens. After the 50th code the message

LIST FULL

will be printed, and the program will continue without asking for any more codes.

When the user enters the range of feature codes to be searched, Q14 will be repeated if the user specifies a lower limit less than 1.0, and Q15 will be repeated if the upper limit is less than the lower limit.

6. Use of the "window" facility (questions 4-7)

The program will operate on any regular series of rectangular map sheets (i.e. the sheets must form a regular rectangular, but not necessarily square, mosaic). The user is allowed to specify a rectangular window in terms of the basic sheet size. Any data falling within the window will be drawn, whilst data outside the window will be omitted. Lines which cross the window boundary will be turned on or off at the appropriate points.

Figure 1 illustrates a case in which 9 map sheets are to be juxtaposed but only the centre sheet and half of the surrounding sheets are required. In this case the user may specify the window as having the following boundaries:

minimum X (i.e. left hand)	0.5 sheets
maximum X (i.e. right hand)	2.5 sheets
minimum Y (i.e. lower)	0.5 sheets
maximum Y (i.e. upper)	2.5 sheets

The user also has the option to have the sheets drawn not in their true position in the mosaic, but with the SW corner of the window moved to the bottom left corner of the drawing paper (or screen, or film, as the case may be). This shifting of the window replaces the "off-set" facility provided in the first version of the drawing program.

Figure 2 illustrates a case in which a single map sheet is drawn, not juxtaposed, and with only the eastern half of the sheet selected. In this case the window specification would be:

minimum X (i.e. left hand)	0.5 sheets
maximum X (i.e. right hand)	1.0 sheets (or any number greater than 1.0)
minimum Y (i.e. lower)	0.0 sheets (or any number less than 0.0)
maximum Y (i.e. upper)	1.0 sheets (or any number greater than 1.0)

The program will also operate on a series of map sheets where the sheets are rectangular but not of equal size. It is only necessary to remember that the window size specification must be given in terms of the size of the first sheet to be drawn.

At a later date it would be possible to extend the window specification facilities so that absolute co-ordinates in units of latitude and longitude or BNG could be requested. It may be useful to add the option to draw data which falls outside the window but not inside, for edge-match checking purposes.

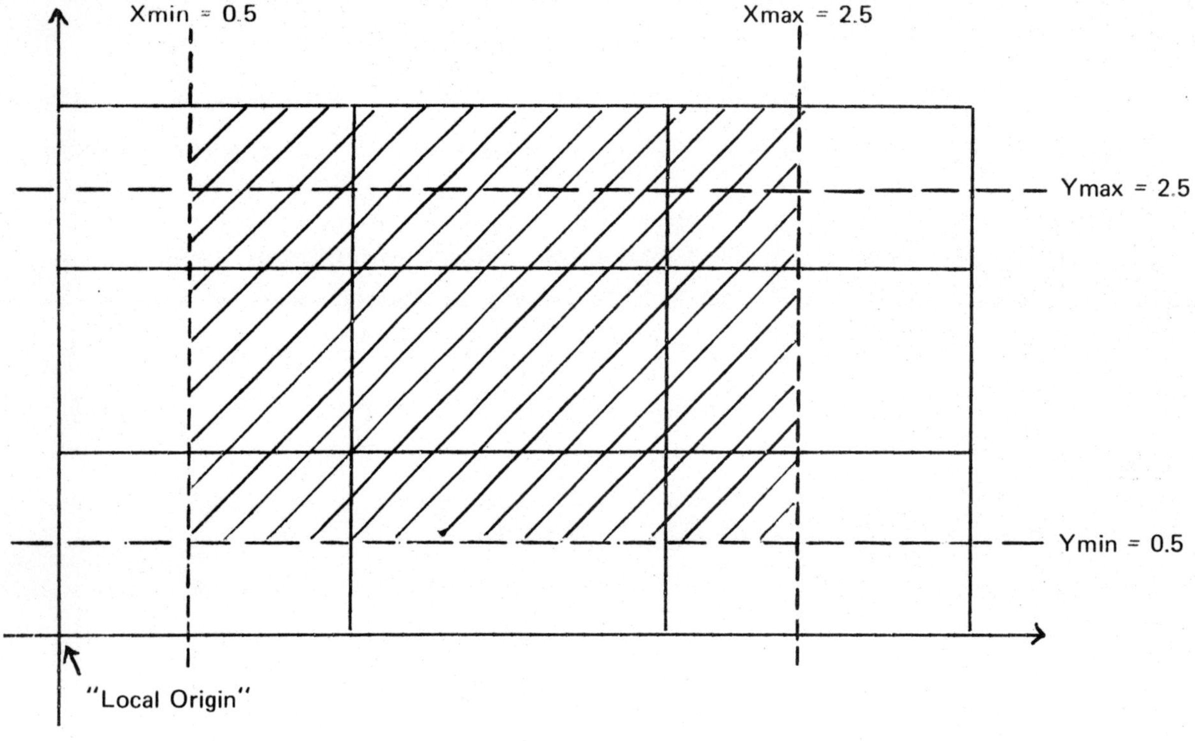

SELECTION OF CENTRAL PORTION FROM MOSAIC OF 9 SHEETS.

FIGURE 1

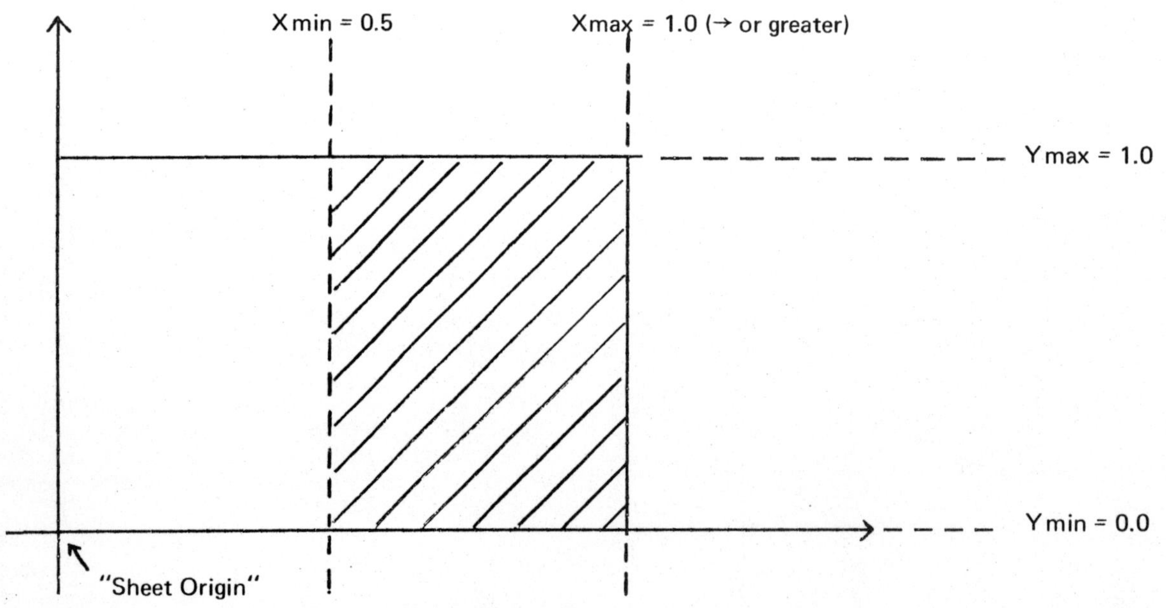

SELECTION OF EASTERN HALF OF SINGLE SHEET

FIGURE 2

Operational Instructions - NGEAL

Program name	NGEAL
Function	To draw on the Gealux (light spot projector) maps stored on IBM tape in standard internal ECU format.
Written by	T. Lang
Modified by	D. Cranch

1. Introduction

This program performs essentially the same functions as NGEAG, except that the map is drawn on the Gealux rather than the Geagraph.

Lines with different feature codes may be drawn with different symbols. Single points may be selected individually according to their feature codes and plotted at a nominated angle from their normal positions.

2. Operation of the Gealux

2.1 Symbol selection

Drawing takes place with one of 50 symbols placed round a symbol disc. It takes one second for a full rotation of the disc, i.e. an average of $\frac{1}{2}$ second to change symbols.

A list of the symbols for the current symbol disc is given in Appendix 1. It is fairly easy to change symbol discs.

2.2 Symbol size

There are two discrete symbol sizes, in the proportion of 1:1:5, with no intermediate variations. A change in magnification is achieved by changing a lens on the projector and raising or lowering the projector to a new position as indicated on a vernier scale. It is possible to produce a single drawing employing both magnifications, but this must be achieved by using two separate passes of the data, one for each magnification.

2.3 Pecked line pattern

Pecked lines may be produced in one of 9 patterns. Which of these patterns is selected may be set by the program. A list of the available patterns is given in Appendix II.

Each pattern is made up of 16 increments, each of which may be on or off (exposed or blank). The patterns are specified by the wiring on a printed circuit electronic card and cannot easily be changed.

The length of each increment (and hence of each total pattern size) can be varied between about 0.18 and 1.2 millimetres. This can not be achieved by program; it is necessary to change a potentiometer setting on the control panel. Therefore it is again necessary to employ a separate pass of the data for each pecking interval that may be required. A graph of pecking interval against potentiometer setting will be found in the Gealux manual.

There is some restriction on the range of drawing speeds which may be used with different pecking intervals. This is described in full in the Gealux manual. There should be no problem with normal curved lines (e.g. contours). For long straight lines with small pecking intervals it will be necessary to turn the speed control knob on the control panel down to about $\frac{1}{2}$ speed.

2.4 Exposure setting

The exposure of lines may be set by program. The variation is on a scale between 1 (bright) and 49 (dim). Suitable values are selected by NGEAL. The variation of light intensity with drawing speed, in order to maintain a constant exposure, is performed by the hardware. At the start and end of the line the drawing speed is zero: to ensure that some exposure takes place at these positions the symbol is additionally projected as if it were a simple point symbol. These point exposures are termed pre and post exposures respectively.

Point exposures can not be controlled by program. A point symbol is actually exposed 3 times, once for the normal point exposure, but also for the pre and post exposures also given to lines.

There are 4 potentiometers on the Gealux control panel which may be used to control the exposure:

(a) Pre-exposure
(b) Point exposure
(c) Post-exposure
(d) Film speed

The effects of these potentiometers are detailed in the Gealux manual. The only potentiometer which normally will require changing is the film-speed control. This potentiometer scales all exposures (point and line) up or down to suit the film being used. A note of the standard settings will be found in the Gealux manual.

3. Program setup

The Gealux should first be set up with the correct magnification, pecking interval and film speed settings. The NGEAL program behaves exactly as NGEAG except that the specification of drawing pens is replaced by a specification of plotting symbols. Questions Q2 to Q17 of NGEAG are replaced by Q100 to Q113 as specified below:-

Q100 ENTER SYMBOL NO. FOR CORNER POINTS
R100 30.

The symbol number is the address of the symbol on the symbol disc. Enter -1. to omit the corner points. (This will be useful on a second pass through the same data but with a different magnification or pecking interval).

Q101 ENTER INDIVIDUAL POINT SYMBOL NO.
R101 24.

Reply -1. to omit all individual points in which case the following questions are omitted and the program skips to Q107.

Q101 is repeated if the reply is not -1. or in the range Ø. to 49 inclusive.

Q102 ENTER ANGLE (DEG.) FROM UPRIGHT
R102 Ø.

The reply determines the angle through which the symbol will be rotated from its usual position before exposure. Enter Ø. for the usual position, 18Ø. for upside-down, and so on.

Q102 will be repeated if the reply is not in the range -36Ø. to 36Ø. exclusive.

Q103 ENTER CODES
R103 30051.
16.
99998.

Any points with the listed feature codes will be played back with the symbol specified in reply to Q101 and Q102. If the list is terminated by 99998. it is assumed that the user wishes to specify further point symbols, and the program repeats from Q101. If the list is terminated by 99999. it is assumed that all specifications have been made.

A maximum of 50 feature codes, all if desired with different symbols and angles, may be selected. If an empty list of codes is specified (i.e. R104 consists of just 99999.) then all individual points will be played back with the same symbol.

If all points have been so selected the program passes to Q107.

If only specific feature codes have been selected, the program asks

Q104 SEARCH CARRIED OUT FROM FEATURE CODE -
R104 2.

Q105 TO FEATURE CODE -
R105 4.

This means that the list of feature codes for each individual point will be searched only from the 2nd to 4th codes to see if a code occurs which has been selected for plotting.

Q106 DO YOU WANT THESE CODES TO BE DRAWN
R106 YES

If the reply is YES, points labelled with the codes specified by R103 will be plotted and all other points omitted. If the reply is NO the reverse will happen (i.e. points with the selected codes will be omitted, and all the rest plotted). If more than one point symbol has been requested, then the selected codes must be plotted, and this question is omitted.

Q107 ENTER LINE SYMBOL NO.
R107 8.

If the reply is -1. the program will omit all lines and begin plotting immediately.

Q108 ENTER DOTTED LINE NO.
R108 3.

The reply should be 0. for a continuous line, 1. to 9. for pecked lines.

Q109 ENTER CODES
R109 50041.
40004.
3.
2.
99998.

Any lines with the listed feature codes will be played back with the symbol specified in reply to Q107 and Q108. If the list is terminated by 99998. it is assumed that the user wishes to specify further line symbols, and the program repeats from Q107. If the list is terminated by 99999. it is assumed that all specifications have been made.

A maximum of 50 feature codes, all if necessary with different symbols and pecking patterns, may be selected. If an empty list of codes is specified (i.e. R109 consists of just 99999.) then all lines are played back, with the same line symbol. If all lines have been selected, the program is ready to plot.

If only specific feature codes have been selected, the program goes on to ask

Q110 SEARCH CARRIED OUT FROM FEATURE CODE -
R110 2.
Q111 TO FEATURE CODE -
R111 4.

This means that the list of feature codes for each item will only be searched from the 2nd to the 4th codes to see if a code occurs which has been selected for plotting.

Q112 DO YOU WANT THESE CODES TO BE DRAWN
R112 YES

If the reply is YES, lines labelled with the codes specified by R109 will be plotted and all other lines omitted. If the reply is NO the reverse will take place (i.e. lines with selected codes will be omitted, and all the rest plotted). If more than one line symbol has been requested, then the selected codes must be plotted and this question is omitted.

Q113 ARE INTERNAL LINES TO BE SUPPRESSED
R113 YES

If the reply is YES, then lines having more than one of the selected feature codes are also omitted (e.g. so that chalk boundaries and clay boundaries may be drawn but not chalk/clay boundaries).

After Q113 the program starts plotting.

4. Error messages

If at any time between Q100 and Q112 the user notices that he has made an error (e.g. specified a wrong symbol number), he may restart the whole selection process from Q100 by typing ↑T on the teletype.

Questions Q100 and Q101 will be repeated if a symbol other than 0. to 49. or -1. is specified. Question Q108 will be repeated if a line pattern outside the range 0. to 9. is specified.

If in reply to Q109 the user selects a feature code outside the range ± 99999. an error message

CODE ERROR
RE-ENTER CODE

will be given. If a code is duplicated, a message will be given

CODE DUPLICATED
RE-ENTER CODE

Question 100 will be repeated if the lower limit of the feature code search is outside the range 1. to 50. Q111 will be repeated if the upper limit is less than the lower limit or greater than 50.

APPENDIX 1. SYMBOL SELECTION (DISC 1) - (November 1969)

GEAGRT/ GEAGRS	GEASYM	SYMBOL NUMBER	SYMBOL	DIMENSIONS IN MMS
␣	32	-0	␣	
0	48	45	0	
1	49	36	1	
2	50	37	2	
3	51	38	3	
4	52	39	4	
5	53	40	5	
6	54	41	6	
7	55	42	7	
8	56	43	8	
9	57	44	9	
A	1	1	|	0.1x2.0
B	2	2	|	0.1x1.0
C	3	3	|	0.1x0.5
D	4	4	|	0.1x0.4
E	5	5	|	0.1x0.2
F	6	6	|	0.1x0.1
G	7	7	|	0.1x0.05
H	8	8	¦	0.1x0.4x1.1
I	9	9	¦	0.1x0.2x0.4
J	10	10	¦	0.1x0.1x0.2
K	11	11	¦	0.1x0.3x0.3
L	12	12	○	2.2
M	13	13	○	1.05
N	14	14	○	0.5
O	15	15	○	0.4
P	16	16	○	0.2
Q	17	17	○	0.1
R	18	18	○	2.8
S	19	19	○	2.4
T	20	20	□	2.0
U	21	21	□	1.0
V	22	22	□	2.6
W	23	23	□	2.2
X	24	24	△	2.5
Y	25	25	△	1.3
Z	26	26	△	3.8
+	43	27	△	3.0
-	45	28	⬡	2.4
.	46	29	⬡ (with dot)	3.1
,	44	30	+	2.0
/	47	31	>	4.0
=	61	32	>	2.5

(continued.........)

Appendix 1 (continued)

GEAGRT/ GEAGRS	GEASYM	SYMBOL NUMBER	SYMBOL	DIMENSIONS IN MMS
(	40	33	-	2.5
)	41	34	-	2.5
*	42	35	∇	2.0
$	36	46	Spare	
;	59	47	Spare	
:	58	48	Spare (Double size)	

APPENDIX 2 - PECKED LINE PATTERNS

STRICHLINIENGENERATOR RCA

(T= 100 x PECK NO.)

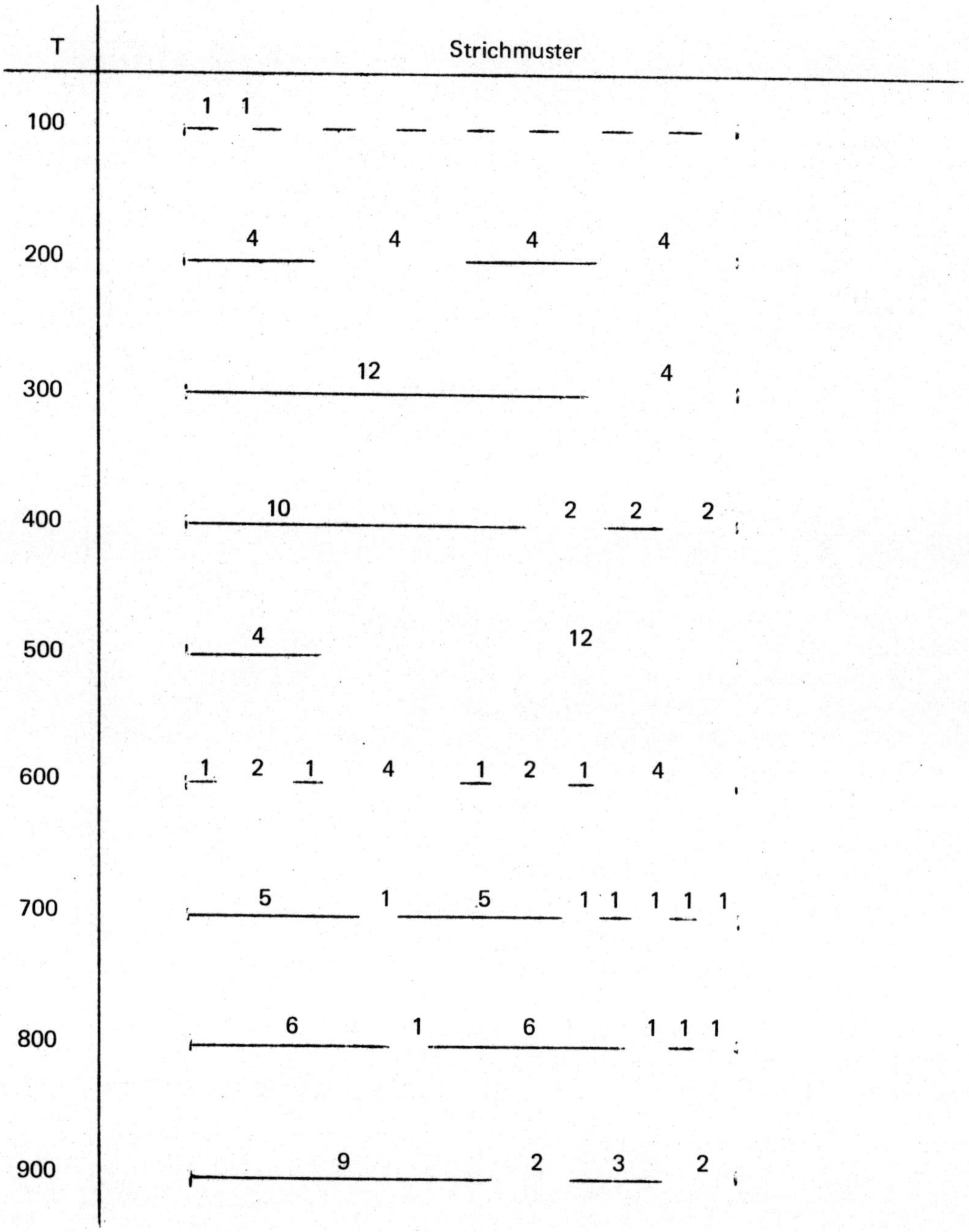

Operational Instructions - NGEANA

Program name NGEANA

Function To draw anaglyphs on the Geagraph from maps stored on IBM tape in standard internal ECU format.

Written by D. Cranch

1. Introduction

The left and right eye views need not be displaced horizontally, but can be displaced at any nominated angle from the horizontal. The program ignores individual points, but no corner points.

2. Program operation

The IBM data tape should be loaded on MT5. (write disabled), and the Geagraph should be set up ready to draw or scribe.

Q1 ENTER SHEET NO.
R1 2.

If the reply is Ø. then it is assumed that all sheets are to be drawn and the program jumps to Q5.

Q2 SAME SPEC AS LAST TIME
R2 NO.

This question is only asked if the answer to Q3 was YES for the previous sheet. If the reply to Q2 is YES, then the new sheet is drawn to the same specification as the previous one, all the following questions being omitted.

Q3 JUXTAPOSE SHEETS
R3 NO.

If the reply is NO the sheet is drawn with its origin at the zero point on the Geagraph table. If the reply is YES, then the sheet is displaced to its correct position relative to the origin for the whole series of sheets.

Q4 ENTER X, Y OFF-SETS IN CMS.
R4 20.
30.

This question is not asked if the sheets are not being juxtaposed. The X, Y off-sets are subtracted from each X, Y co-ordinate pair prior to plotting, and are used to ensure that the sheets being plotted are in the drawing area of the table.

Q5 ENTER SCALE
R5
(a) N2.Ø for twice the Nominal scale of the sheet
(b) DØ.25 for one-quarter of the scale of the digitised sheet
(c) X20.
Y1Ø. for a drawn map of size 20cm. x 10cm.

Q6 ENTER TOLERANCE IN MMS
R6 0.1

Q7 ENTER CODE SELECTION
END with 99999.
R7 1000.
2000.
3000.
4000.
5000.
99999.

The operator must enter here the feature code for each contour which is required to be plotted, terminating the list with 99999.

Q8 SEARCH FROM FEATURE CODE NO.
R8 1.

Q9 TO FEATURE CODE NO:
R9 1.

This tells the program to look at only the first feature code of each line when comparing with the list entered at R7.

Q10 DO YOU WANT THESE LINES TO BE DRAWN
R10 YES

This reply causes lines having one of the entered codes to be drawn; to reply NO would have the effect of suppressing these lines whilst drawing all other lines.

Q11 DO YOU WANT INTERNAL LINES SUPPRESSED
R11 YES

Lines with more than one code in the entered list will be omitted unless the reply NO is given.

Q12 WHICH FEATURE CODE IS HEIGHT
R12 1.

The program will take the first code of a chosen line item as a contour height.

Q13 ENTER NO. OF VERTICAL UNITS TO EQUAL 1MM. DISPLACEMENT
R13 1000.

The reply to this question determines the vertical scale of the anaglyph.

Q14 ENTER FEATURE CODE VALUE OF TABLE LEVEL CONTOUR
R14 5ØØØ

The apparent height of any contour can be obtained from:

$$\text{apparent height (mm.)} = \frac{\text{(value of feature code of the contour) - R14}}{\text{R13}}$$

Q15 ENTER ANGLE (DEG) OF EYES FROM HORIZONTAL
R15 90°

The reply to this question determines at what angle, clockwise from the horizontal, the viewer's head must be turned to obtain the stereoscopic effect.

3. Program running

When the reply to Q15 has been given, the program will type the message:

PEN FOR LEFT EYE VIEW
PAUSE

When ↑P is typed the program will drive the Geagraph and draw the left eye's view; when this is completed the program will type:

CHANGE PEN
PAUSE

The appropriate response here is to change the Rapidograph pen (or Masonscribe sheet) and type ↑P.

The program will then draw the right eye's view and type the above message again, i.e.

CHANGE PEN
PAUSE

If Ø was entered at R1, the program will start plotting the left eye's view of the next required sheet upon typing ↑P, otherwise it will print the following message:

DRAWING TOOK 3 MINS 50 SECS

and then ask

ANY MORE SHEETS FROM THIS TAPE

If the reply is YES the program loops to Q1, otherwise it asks:

ANY MORE TAPES

If the reply to this question is YES, the program asks for a new tape to be loaded on MT5.

LOAD NEW TAPE
PAUSE

and loops to Q1 when ↑P is typed.

Operational Instructions - NJOIN

Program name NJOIN (MKIA)

Function To read an IBM tape in standard system format, rotate and translate the map co-ordinates so that the south west corner is at the origin with the south border along the X axis, and calculate X and Y offsets relative to a convenient local origin (usually the extreme south west corner of the set of maps).

Written by D. Vicars

1. Introduction

This program is only applicable to map sheets whose corners are referenced in geographical coordinates in one of the following projections:

(a)	Gall	(GALL)
(b)	Modified Gall	(MGALL)
(c)	Mercator	(MERC)
(d)	Equidistant cylindrical	(EDCYL)
(e)	Equal area cylindrical	(EACYL)
(f)	Orthomorphic Zenithal	(OZEN)
(g)	Equidistant Zenithal	(EDZEN)
(h)	Equal area Zenithal	(EAZEN)
(i)	Orthomorphic Conical	(OCON)

(This is the lambert, with two standard parallels at 52° 56 N and 36° 15 N, by the French IGN for 1 : 1,000,000 maps)

The abbreviations in brackets are the replies which are recognised for question 7.

The program provides the facility to merge several imput files, e.g. for combining two or more parts of a map that takes up more than 1 D-mac tape.

2. Program setup

Before execution the IBM tape units must be set up as follows: Input tape on MT5 write disabled, output tape on MT6 write enabled. The program begins by typing its name, version number and date of issue. It then asks a series of questions -

Q1 IS THIS THE FIRST FILE ON O/P TAPE
R1 YES

If the reply is "NO" then Q2 is omitted and the output tape spaced forward to just before the end of the data already on it.

Q2 ENTER TAPE NUMBER
R2 MT 122

The tape number should be given as either "MT" or "DT" followed by the number terminated by a decimal point (all numbers must be terminated by a decimal point).

Q3 ENTER INPUT FILE NUMBER
R3 5.

The input tape is positioned at file 5. If the tape does not have 5 files, the program types "FILE 5 NOT FOUND" and repeats Q3.

Q4 ENTER DATE AS DD-MM-YY
R4 22-Ø1-7Ø

The reply is checked and if any number is out of range (e.g. the month greater than 12) then "ERROR" is typed and the question repeated.

Q5 ENTER YOUR INITIALS
R5 A.K. (e.g.)

Q4 and Q5 are only asked once each run.

Q6 ENTER RUN DESCRIPTION
R6 JOIN OF FILES 5, 6 & 7.

The first 25 characters are stored as a comment in the file description block.

Q7 WHICH PROJECTION
R7 MERC

The reply must be one of those listed in the introduction otherwise the program types "REPLY UNKNOWN" and repeats the question.

The program now asks the questions necessary to calculate the transform and offsets for all coordinates, they are only asked once for each output file.

Q8 TYPE THE CORNER POINT CO-ORDINATES SW CORNER -
R8 N48
E17

This is repeated for the other three corners (NW, NE and SE) after which they are printed out for inspection and question 9 asked.

Q9 ARE THEY OK
R9 YES

If the reply is "NO" then Q8 is repeated.

Q10 TYPE SCALE
R10 1ØØØØØØ

The number entered is taken to be the reciprocal of the scale (e.g. 2500 for 1/2500)

Q11 TYPE LOCAL ORIGIN COORDS
R11 N4Ø.
E1Ø.

Q12 DOES THE TAPE HAVE 4 CORRECT CALIBRATION POINTS
R12 YES

This causes the program to skip questions Q13 to Q16.

The operator has the opportunity to delete and re-order the corner points if the reply is "NO".

Q13 HOW MANY CALIBRATION POINTS
R13 6.

If the number is less than 4 or greater than 9, the question is repeated. If the number is greater than 4 then question 14 is asked:

Q14 2 POINTS MUST BE DELETED, TYPE THEIR NUMBERS
R14 2.
R14 4.

If a number is repeated or greater than the total number of calibration points then the message "ERROR IN LAST NUMBER, TYPE IT AGAIN" is typed and the user should reply with the correct number.

Q15 WHICH CORNER IS CALIBRATION POINT n
R15 SW

The question is asked 4 times with n taking the values of the remaining calibration point numbers (e.g. 1, 3, 5 and 6). The replies must be one of "SW", "NW", "NE" or "SE". If the reply is not one of these the question will be repeated with the same value of n. If a reply has already been used it will be rejected with the comment: "SW CORNER ALREADY USED" and the question repeated. When this has finished a summary is printed followed by question 16.

Q16 IS THIS OK
R16 YES

If the reply were "NO" then the process is repeated from Q14 to enable the operator to correct mistakes. The input file is then read and if the number of corner points is different from the number claimed the message "5 CORNER POINTS FOUND INSTEAD OF 6" is typed and the program skips to question Q22 to allow the operator to select another file.

The corner points read from tape are first checked for separation and if the distance between any two is small the message:

"CORNERS 1 AND 2 ARE CLOSER THAN 10MM.
DISTANCE = Ø.ØØØ1 MM."

and the program skips to question Q18 so that the operator can select another file or have another attempt at the current file.

If all the corners are more than 1 cm apart then the corners are checked for congruency and similarity with the given shape, giving one of the following messages if either of the checks fail.

(i) "THE MAP IS A DISSIMILAR SHAPE FROM THE ONE YOU HAVE GIVEN"

or

(ii) "THE MAP IS NOT ON THE STATED SCALE
E-W SCALE FACTOR = Ø.875Ø N-S SCALE FACTOR = Ø.8511"

This is followed by

Q17 DO YOU WANT TO START AGAIN
R17 NO

This causes the program either to attempt to fit the dissimilar shape or to output the map at the nominal scale with the scale factors stored for future use by NGEAG or AMALG. The reply "YES" allows the operator to re-order the corner points or select another file; the program skips to question 22.

The program processes all the input file applying the transformation to every point. When the end of the input file is read the program types "END OF I/P FILE" followed by

Q18 IS THERE ANOTHER FILE
R18 NO

This causes the program to take the appropriate action necessary to ensure a correctly formatted output tape. The program then skips to Question 21. If the reply was "YES" then the program asks:

Q19 ENTER INPUT FILE NUMBER
R19 93

If an output file is already open (e.g. file 93. is the second or subsequent input file for the current output file) then the program checks the descriptive information in its header block against the corresponding data in the first file and if they agree the program goes to Q24 before merging the file into the current output file. Otherwise the details are printed and the operator has the chance to accept or reject the file.

Q20 THIS FILE IS FOR N10 GULF OF ADEN SHEET 1.
DO YOU WANT IT
R20 YES

This accepts the file and the program returns to question 18 when it reaches the end of the file.

The reply "NO" rejects the file and the program repeats question 19.

If an output file has to be closed by the reply to question 18 then the program asks:

Q21 DO YOU WANT TO START ANOTHER FILE
R21 YES

The operator now has the chance to open another tape. If the reply is "NO" then the program will close the output tape and stop the run.

Q22 DO YOU WANT TO OPEN ANOTHER TAPE
R22 NO

This causes the program to skip to Q3. If the reply is "YES" then the program pauses to allow another tape to be mounted and when it is restarted(by "CTRL" and "P") or rewinds the tape before continuing with Q3.

3. Error messages

IOPS4 A magnetic tape unit is not ready, e.g. the output tape has no write permit rung or is not "on line". The program can be restarted by typing "CTRL" and "R" together once the device has been made ready.

Operational Instructions - NMTTDT

Program name NMTTDT

Function To read an IBM tape produced on the D-mac digitiser, convert the data from D-mac format to standard system format, and prepare a corresponding output IBM tape (which may then be used as input to the NJOIN program).

Written by T. Lang

1. Introduction

This program performs essentially the same process as MTTDT in the old system, with the following major alterations:-

a) Multiple feature codes are allowed

b) Output is now to a second IBM tape, rather than DEC tape

c) Correction data, intended for subsequent input to CORECT, is checked for consistency.

2. Input of Multiple Feature Codes on the D-mac

Up to 50 codes may be entered for each number.

During the digitising process each code must be prefixed by the use of the "KEYBOARD" button on the D-mac control panel.

```
e.g.  &lt;K&gt; 123
      &lt;K&gt; 89
         :
      line data
         :
```

will associate codes 123 and 89 with the line which follows. A code may consist of any number of digits, providing it does not exceed 131,071. A code may be prefixed by a minus sign to denote negative quantities.

To erase a code which has just been entered, a "/" character may be used.

e.g. <K> 161

<K> 32/

<K> -32

⋮

line information

⋮

will associate codes 161 and -32 with the line which follows.

If no new codes are entered, the current set of feature codes will be remembered, unless the answer to Q14 is NO, and carried on for subsequent items

e.g. <K> 161

<K> -32

<K> 94

⋮

line 1

⋮

line 2

⋮

will associate codes 161, -32 and 94 with both lines 1 and 2. If new codes are entered then they will replace the codes entered previously, starting with the first code in the list:-

e.g. <K> 161

<K> -32

⋮

line 1

⋮

<K> 99

⋮

line 2

⋮

<K> 64

⋮

line 3

⋮

will result in the feature code assignments

line 1	161	-32
line 2	99	-32
line 3	64	-32

To erase entirely the memory of previous feature codes a "x" character may be used

e.g. <K> 161

<K> -32

line 1

<K> x

<K> 99

line 2

<K> 64

line 3

will result in the feature code assignments

line 1	161	-32
line 2	99	
line 3	64	

3. Input of Correction Codes on the D-mac

If a D-mac tape is to form the correction data for the program CORECT, then every item must be prefixed by a pair of alphabetic characters denoting the correction to be applied. The correction characters are entered exactly as if they were feature codes.

e.g. <K> IL

<K> 161

line 1

<K> DL

point 1

(A full description of the use of correction codes, and a complete list of all permitted codes, will be found in the operational instructions for the CORECT program.)

4. Program setup

Before program execution is initiated, the IBM tape units should be set up as follows:-

MT5 Input D-mac tape (write disabled).

MT6 Output tape (write enabled)

The program will start by asking a series of questions, which are best illustrated by example

Q1 IS THIS THE FIRST FILE ON O/P TAPE
R1 YES

If the reply is NO, question 2 will be omitted

Q2 ENTER TAPE NUMBER
R2 MT34.

The serial number of the tape is recorded on the tape itself. The tape number should be supplied as MT (for IBM tape) or DT (for DEC tapes), followed by the tape number, as written on the label stuck to the tape spool, and terminated by a decimal point.

Q3 WHICH DMAC SHEET (START AT 1.0)
R3 13.

The reply indicates which sheet from the D-mac tape is to be processed, and is given as a floating point number. If the requested sheet can not be found an error message will be given - see section 5. Otherwise, having located the requested sheet, the program asks for information describing the map to be processed.

Questions 4 and 5 are asked only once, and the replies carried over if more than one D-mac sheet is processed.

Q4 ENTER DATE AS DD-MM-YY.
R4 12-01-70.

Q5 ENTER OPERATORS INITIALS (5 CHARS)
R5 AK

Questions 6 to 9 are intended to obtain a specification of the map being processed.

Q6 ENTER PROJECT NAME (5 CHARS)
R6 N103.

Q7 ENTER GENERAL PROJECT DESCRIPTION (25 CHARS)
R7 TRACK DATA FOR CONTOURING.

Any number of characters from 0 to 25 may be supplied. The description entered here is a permanent title which will always be associated with the data however it is processed.

Q8 IS THIS A CORRECTION SHEET
R8 NO

If the reply is YES, it is assumed that the sheet contains correction data for the CORECT program. Every item will then be checked for a valid correction code (see section 7).

Q9 ENTER FILTER DISTANCE IN MMS.
R9 0.25

The program filters line data so that no two points on any line are closer than the nominated distance.

If a second or further D-mac sheet is processed, question 6 is prefixed by

Q6a SAME SPEC AS LAST SHEET
R6a YES

If the reply is YES, the details elicited by questions 6 to 9 are carried over for the new sheet, and the program proceeds immediately to question 10.

If the reply is NO, questions 6 to 9 are repeated.

Q10 ENTER ANY SPECIAL COMMENTS FOR THIS SHEET (25 CHARS)
R10 REPEAT OF RUN ON 10/01/70.

This title is used only to label the current data, and is overwritten when further processing - e.g. JOIN - is applied. These comments are intended only to draw attention to special circumstances, and will generally be left blank.

Q11 ENTER OVERALL MAP REFERENCE (10 CHARS)
R11 N10 231/A

Q12 ENTER SHEET NUMBER (E.g. 5.0)
R12 4.0

Q13 ENTER MAXIMUM NUMBER OF FEATURE CODES (E.g. 1.0)
R13 2.0

This specifies the maximum number of codes expected for each item. If more than the expected number of codes appears, then all the extra codes are used in turn to overwrite the last legitimate code. E.g. if the reply to Q13 was 2.0 then

```
<K> 14
<K> 27
<K> 26    (overwrites 2nd code)
<K> 32    (also overwrites 2nd code)
    |
   line
    |
    |
```

would result in the line being given codes 14 and 32.

Q14 ARE FEATURE CODES TO BE CARRIED OVER
R14 YES

If the reply is YES the program carries over feature codes from one point or line to the next unless specific instructions to the contrary are given. If the reply is NO then all memory of previous feature codes is erased before each item is processed. In the latter circumstances however, in the exceptional cases where codes are to be carried over, a '#' character may be used.

```
e.g. <K> 123
     <K> -31
     <K> 164
     <K>   3
      |
     line 1
      |
     <K>#
     <K> 14
      |
     line 2
      |
     <K> 9041
      |
     line 3
      |
      |
```

would result in the feature code assignments

line 1	123	-31	164	3
line 2	14	-31	164	3
line 3	9041			

(i.e. * clears the feature code memory when codes are normally carried over, # retains the feature code memory when codes are normally not carried over).

5. Program termination

A number of messages indicate that the end of the current sheet has been reached.

END OF SHEET

This will be output if the sheet was found to terminate normally (i.e. "END" button was used correctly on the D-mac).

SHEET TERMINATED BY CORNER POINT OF NEXT SHEET

Where the D-mac operator forgot to use the "END" button, and next sheet follows immediately.

SHEET TERMINATED BY END OF FILE OR END TAPE

Where the D-mac operator forgot to use the "END" button, and this was the last sheet on the tape.

The program will then go on to ascertain if there are any more D-mac sheets to be processed, by asking the questions

Q15 ANY MORE SHEETS ON THIS DMAC TAPE
R15 NO

If the response is YES, the program loops back to question 3.

Q16 ANY MORE DMAC TAPES
R16 NO

If the reply is YES the program will type

LOAD NEXT DMAC TAPE
PAUSE

After the new tape is loaded, type ↑P on the keyboard to continue. The program loops back to Q3.

6. Error Conditions

If the input tape unit (i.e. with the D-mac tape) is not set to unit number 5, or not switched "on-line" the program will loop waiting for these conditions to be satisfied, but no message will be printed.

If a parity error is detected on the input tape, the program will loop, alternately back-spacing over the bad record and attempting to read it. (This has never happened in practice.)

If the output tape unit is not set to unit no. 6, or not switched "on-line" the program will type "IOPS 4". After resetting the unit correctly, type ↑R on the teletype to continue.

Co-ordinates are stored on the D-mac tape as 4 digit numbers. In the event of a hardware error a digit may be corrupted (or omitted).

If in response to Question 3 the user gives the number of a D-mac sheet greater than the total number of sheets on the tape, the error message

END FILE/TAPE ON DMAC TAPE

will be given, and the program will then loop to question 15.

7. Correction code checking

If the D-mac sheet is a correction sheet, then every item will be checked to see that it has a valid correction code associated with it. As each item is processed the corresponding correction code will be listed on the teletype. This is intended as a check that all corrections were included, and also to assist recovery in the event of an error being detected.

e.g.

No.	Correction
1	DL
2	DL
3	IL
4	IL
5	CL
6	CP

If an error is detected one of the following error messages will be given

a) INVALID CORRECTION INSTRUCTION JL
Where JL is the invalid correction code

b) POINT FOLLOWS LINE CORRECTION INSTRUCTION IL
Where a point has a correction appropriate to a line associated with it

c) LINE FOLLOWS POINT CORRECTION INSTRUCTION DL
Where a line has a correction appropriate to a point associated with it

d) MISSING CORRECTION INSTRUCTION ASSOCIATED WITH POINT

e) MISSING CORRECTION INSTRUCTION ASSOCIATED WITH LINE
After any of these errors, the user will be asked to supply the correct correction instruction

Q ENTER NEW INSTRUCTION (2 CHARS)
R DL (following error (b) for example).

8. Detection of D-mac errors

If an illegal digit is detected in a coordinate, the entire item is omitted (i.e. not copied to the output tape). If the faulty item is a point within a line no harm will be done. If the faulty item is an individual point or corner point, serious corrective action will be required at a later stage (e.g. in JOIN or CORECT programs).

The error message is

INVALID CHAR IN DMAC NUMBER 12

Which will be followed by one of

IN LINE
IN POINT
IN CORNER

The program will not PAUSE after the detection of such an error, but will continue immediately with the next item on the tape.

Operational Instructions - NSD

Program name NSD

Function To draw a map file on the storage display

Written by T. Lang

1. Introduction

This program reproduces map files on the storage display. It offers exactly the same facilities as NGEAG, and operates with the same questions and responses. (The sole exception being that no request should be made to change drawing pens.)

Before the program is called the display should be switched on and cleared by pressing the "erase" button. The program issues no "erase" commands of its own, and so successive maps will be added to the display as they are processed. The user may push "erase" between map sheets to erase previous material from the screen.

2. Program setup

Same as for NGEAG, except that a storage tube is used instead of the Geagraph.

Operational Instructions - NSJOIN

Program name NSJOIN MK1B

Function To read an IBM tape in standard system format, rotate and translate the map co-ordinates so that the south west corner is at the origin with the south border along the X axis, and calculate X and Y offsets relative to a convenient local origin (usually the extreme south west corner of the set of maps).

Written by D. Vicars

1. Introduction

This program positions a sheet so that its south west corner is at the origin and the south and west borders lie along the X and Y axes respectively. At the same time small distortions are removed and positional information and scale are stored in the header block. Information recorded on the D-mac must be passed through these processes before it can be utilised properly by the system.

2. Program setup

Before execution the IBM tape units must be set up as follows: Input tape on MT5 write disabled, output tape on MT6 write enabled. The program begins by typing its name, version number and date of issue. It then asks a series of questions -

Q1 IS THIS THE FIRST FILE ON O/P TAPE
R1 YES

If the reply is "NO" then Q2 is omitted and the output tape spaced forward to just beyond the end of the data already on it.

Q2 ENTER TAPE NUMBER
R2 MT 122.

The tape number should be given as either "MT" or "DT" followed by the number terminated by a decimal point (all numbers must be terminated by a decimal point).

Q3 ENTER INPUT FILE NUMBER
R3 5.

The input tape is positioned at file 5. If the tape does not have 5 files, the program types "FILE 5 NOT FOUND" and repeats Q3.

Q4 ENTER DATE AS DD-MM-YY
R4 22-Ø1-7Ø

The reply is checked and if any number is out of range (e.g. the month greater than 12) then "ERROR" is typed and the question repeated.

Q5 ENTER YOUR INITIALS
R5 A.K.

Q4 and Q5 are only asked once each run.

Q6 ENTER RUN DESCRIPTION
R6 JOIN OF FILES 5, 6 & 7.

The first 25 characters are stored as a comment in the file description block.

Q7 WHICH PROJECTION
R7 GRID

If R7 is "NON" (for non-standard projections) then the program continues from Q13 and asks some further questions when the corner points have been read. Only these two replies are recognised; any other will cause the question to be repeated.

The program now asks the questions necessary to calculate the transform and offsets for all co-ordinates, they are only asked once for each output file.

Q8 TYPE THE CORNER POINT CO-ORDINATES IN GRID UNITS
EASTINGS THEN NORTHINGS SW CORNER-
R8 242.
129.

This is repeated for the other three corners (NW, NE and SE) after which they are printed out for inspection and question 9 asked

Q9 ARE THEY OK
R9 YES

If the reply is "NO" then Q8 is repeated.

Q10 TYPE SCALE
R10 25ØØ.

The number entered is taken to be the reciprocal of the scale (e.g. 25ØØ. for 1/25ØØ.)

Q11 TYPE LENGTH OF CELL SIDE IN METRES
R11 1ØØØ.

Q12 TYPE LOCAL ORIGIN IN GRID CO-ORDS
EASTINGS THEN NORTHINGS
R12 24Ø.
12Ø.

Q13 DOES THE TAPE HAVE 4 CORRECT CALIBRATION POINTS
R13 YES

This causes the program to skip to question Q18.

The operator has the opportunity to delete and re-order the corner points if the reply is "NO".

Q14 HOW MANY CALIBRATION POINTS
R14 6.

If the number is less than 4 or greater than 9. the question is repeated. If the number is greater than 4 then question 15 is asked:

Q15 2 POINTS MUST BE DELETED, TYPE THEIR NUMBERS
R15 2.
R15 4.

If a number is repeated or greater than the total number of calibration points then the message "ERROR IN LAST NUMBER, TYPE IT AGAIN" is typed, and the user should reply with the correct number.

Q16 WHICH CORNER IS CALIBRATION POINT n
R16 SW

The question is asked 4 times with n taking the values of the remaining calibration point numbers (e.g. 1, 3, 5 and 6). The replies must be one of "SW", "NW", "NE" or "SE". If the reply is not one of these the question will be repeated with the same value of n. If a reply has already been used it will be rejected with the comment: "SW CORNER ALREADY USED" and the question repeated. When this has finished a summary is printed followed by question 17.

Q17 IS THIS OK
R17 YES

If the reply were "NO" then the process is repeated from Q14 to enable the operator to correct mistakes. The input file is then read and if the number of corner points is different from the number claimed the message "5 CORNER POINTS FOUND INSTEAD OF 6" is typed and the program skips to question Q22 to allow the operator to select another file.

If the input map is on a non-standard projection (i.e. the reply to question 7 was "NON") then the following questions are asked

Q18 ENTER SCALE
R18 2ØØØØØØ.

Q19 ENTER X OFFSET IN MMS.
R19 1ØØ.

Q2Ø ENTER Y OFFSET IN MMS.
R2Ø Ø.

The corner points read from tape are first checked for separation and if the distance between any two is small the message:

"CORNERS 1 AND 2 ARE CLOSER THAN 1ØMM.
DISTANCE = Ø.ØØØ1MM."

and the program skips to question Q22 so that the operator can select another file or have another attempt at the current file.

If all the corners are more than 1cm apart then the corners are checked for congruency and similarity with the given shape, giving one of the following messages if either of the checks fail

(i) "THE MAP IS A DISSIMILAR SHAPE FROM THE ONE YOU HAVE GIVEN"

or

(ii) "THE MAP IS NOT ON THE STATED SCALE
E-W SCALE FACTOR = Ø.875Ø N-S SCALE FACTOR = Ø.8511"

This is followed by

Q21 DO YOU WANT TO START AGAIN
R21 NO

This causes the program either to attempt to fit the dissimilar shape or to output the map at the nominal scale with the scale factors stored for future use by NGEAG or AMALG. The reply "YES" allows the operator to re-order the corner points or select another file; the program skips to question 22.

The program processes all the input file applying the transformation to every point. When the end of the input file is read the program types "END OF I/P FILE" followed by

Q22 IS THERE ANOTHER FILE
R22 NO

This causes the program to take the appropriate action necessary to ensure a correctly formatted output tape. The program then skips to Question 25. If the reply was "YES" then the program asks:

Q23 ENTER INPUT FILE NUMBER
R23 93.

If an output file is already open (e.g. file 93. is the second or subsequent input file for the current output file) then the program checks the descriptive information in its header block against the corresponding data in the first file and if they agree the program goes to Q24 before merging the file into the current output file. Otherwise the details are printed and the operator has the chance to accept or reject the file.

Q24 THIS FILE IS FOR IGS GEOPHYSICAL CONTOURS SHEET 1.
DO YOU WANT IT
R24 YES.

This accepts the file and the program returns to question 22 when it reaches the end of the file.

The reply "NO" rejects the file and the program repeats question 23.

If an output file has to be closed by the reply to question 22 then the program asks:

Q25 DO YOU WANT TO START ANOTHER FILE
R25 YES

The operator now has the chance to open another tape. If the reply is "NO" then the program will close the output tape and stop the run.

Q26 DO YOU WANT TO OPEN ANOTHER TAPE
R26 NO.

This causes the program to skip to Q3. If the reply is "YES" then the program pauses to allow another tape to be mounted and when it is restarted (by "CTRL" and "P") it rewinds the tape before continuing with Q3.

3. Error Messages

IOPS4 A magnetic tape unit is not ready, e.g. the output tape has no write permit rung or is not "on line". The program can be restarted b y typing "CTRL" and "R" together once the device has been made ready.

NSJOIN APPENDIX
SAMPLE PRINT OUT

```
 NJOIN MK 1A. DV 27-11-69
 IS THIS THE FIRST FILE ON O/P TAPE

 NO

 ENTER INPUT FILE NO.
>  11.Ø
 ENTER DATE AS DD-MM-YY
> 27-Ø1-7Ø
 ENTER YOUR INITIALS
> AJK
 ENTER RUN DESCRIPTION
> BIDEFORD
 WHICH PROJECTION
> GRID
 TYPE THE CORNER POINT COORDS IN GRID UNITS
 EASTINGS THEN NORTHINGS.
 SW CORNER -
> 244.
> 127.
 NW CORNER -
> 244.
> 128.
 NE CORNER -
> 245.
> 129/8.
 SE CORNER -
> 245.
> 127.
 CORNER POINTS -
 SW - 244E,   127N.
 NW - 244E,   128N.
 NE - 245E,   128N.
 SE - 245E,   127N.  ARE THEY OK
> YES
 TYPE SCALE
> 25ØØ.
 TYPE LENGTH OF CELL SIDE IN METERS
> 1ØØØ.
 TYPE LOCAL ORIGIN IN GRID COORDINATES,
 EASTINGS THEN NORTHINGS
> 24Ø.
> 12Ø.
 DOES THE FILE HAVE 4 CORRECT CALIBRATION POINTS -
> NO
```

```
 HOW MANY CALIBRATION POINTS -
> 5.
 1 POINT MUST BE DELETED, TYPE ITS NUMBER
> 2.
 WHICH CORNER IS CALIBRATION POINT 1
> SW
 WHICH CORNER IS CALIBRATION POINT 3
> NW
 WHICH CORNER IS CALIBRATION POINT 4
> NE
 WHICH CORNER IS CALIBRATION POINT 5
> SE


 SUMMARY
 CALIBRATION POINT 1 IS SW CORNER.
 CALIBRATION POINT 2 DELETED.
 CALIBRATION POINT 3 IS NW CORNER.
 CALIBRATION POINT 4 IS NE  CORNER.
 CALIBRATION POINT 5 IS SE  CORNER.


 IS THIS OK
> YES
 END OF I/P FILE
 IS THERE ANOTHER
> YES
```

Operational Instructions - PROJCH

Program name PROJCH Mk 2A

Function To change the projection of a map held in a single file on magnetic tape

Written by D. Vicars

1. Introduction

The program provides transformations between several standard projections, using spherical formulae, with the option of a pole change on output to obtain skew and transverse projections. When the pole is changed, or if the input is in GRID format (e.g. with grid lines at kilometer intervals) the graticule generating subroutine is invoked and the output is automatically aligned within the graticule.

The input is assumed to have been reduced to a minimum volume by selection and filtering with AMALG.

2. Program setup

The program requires the input tape to be mounted on unit 5, write disabled, and the output on unit 6, write enabled. When loaded the program first asks the questions necessary to update the file description and requests the definition of the output projection and graticule.

Q1 IS THIS THE FIRST FILE ON O/P TAPE
R1 NO

This causes the output tape to be positioned just beyond the data already on it. If the reply was "YES", then question 2 is asked.

Q2 ENTER TAPE NUMBER
R2 MT98.

Q3 ENTER INPUT FILE NUMBER
R3 5.

If the number is less than 1. or greater than 25Ø. then the program will repeat the question. Otherwise, it searches the input tape for the given file. If there are less than 5. files on the tape the program will type 'FILE 5 NOT FOUND' and repeat the question.

Q4 ENTER DATE AS DD-MM-YY
R4 17-Ø3-7Ø

The reply is given some simple checks: if any of them fail then the program types 'ERROR' and repeats the question.

Q5 ENTER YOUR INITIALS
R5 C.W.

Only 5 characters are retained.

Q6 ENTER RUN DESCRIPTION
R6 COASTLINE ONLY

This is to enable this run to be identified more closely than the standard file description permits. 25 characters are retained.

The program will then type a list of the accepted abbreviations for the currently implemented projections:

Input and output

GALL	Gall's projection
MGALL	Modified Gall projection
MERC	Mercator's Projection (Orthomorphic Cylindrical)
EDCYL	Equidistant Cylindrical
EACYL	Equal Area Cylindrical
OZEN	Orthomorphic zenithal
EDZEN	Equidistant zenithal
EAZEN	Equal Area zenithal
OCON	Orthomorphic Conical (This is the projection used by the IGN for 1:1000000 maps of France, with standard parallels at 52°56'N and 36°15'N)

For input only

BNG	British National Grid (basically transverse mercator on 2°W)

The program types the input projection and then asks question 7:

Q7 ENTER OUTPUT PROJECTION
R7 OZEN

The reply must be one of the abbreviations listed above, otherwise the program will type 'NAME UNKNOWN', where X is the first 5 characters of the reply, and repeat the question.

Q8 INPUT SCALE = 1ØØØØØØ.
ENTER OUTPUT SCALE
R8 2ØØØØØØ.

A change in scale may be necessary if the pole is to be changed.

Q9 DO YOU WANT TO CHANGE THE POLE
R9 YES.

This causes the program to ask:

Q10 ENTER COORDS OF NEW POLE: E THEN N
R10 9∅.
∅.

The replies must be in degrees and decimals of a degree.

The user is then asked to define the boundaries of a rectangular area on the table which is to contain the output. The first time any point in the new projection falls outside this area, a warning message will be generated, and the point omitted.

Q11. ENTER BOUNDARY SPECS IN CMS.
Q11a LOWER X
R11a ∅.

Q11b UPPER X
R11b 2∅.

Q11c LOWER Y
R11c ∅.

Q11d UPPER Y
R11d 3∅.

If the input map is in BNG projection, the program will convert the BNG coordinates of the map corner points to geographicals and type them. If the pole of the projection is to be changed the program will type out the geographical coordinates of the corners in the new system. In both these cases the user has the opportunity to invoke the graticule generating subroutine.

Q12 DO YOU WANT A NEW SYSTEM GRATICULE
R12 YES

If the reply is 'NO', then Q13 is skipped and only a border will be output.

Q13 HOW MANY DEGREES BETWEEN LINES
R13 2.

The program will now proceed with the processing. If any point lies outside the boundary defined in Q11 the message 'BOUNDARY

EXCEEDED' will be given on the first occurrence, unless the point is a corner point, in which case the message 'BOUNDARY EXCEEDED BY CORNER n', where n is the number of the corner point, will be given for each corner point outside, and that point ignored. No further message will be given for other points outside the zone.

When the complete file has been processed the program types 'STOP' and returns to the monitor.

Operational Instructions - READFF

Program name READFF

Function Read a binary figure field and write it to the teletype

Written by D.S. Connelly

1. Introduction

It is often convenient to read through a figure field by transferring it to the teletype for visual inspection. This program accomplishes that transfer. The operator can interact with the program to interrupt the printing of the current column vector and skip to the next one.

2. Program setup

Mount the figure field tape with the ring removed on MAGTAPE drive 5. Mount the figure field operations tape on DECTAPE drive 2. Initiate the DEC MONITOR and type E READFF.

The program immediately begins reading and printing the figure field. The first set of numbers to be printed is a parameter header. All the remaining sets of numbers are column vectors. Each set will be printed as a block of 201 numbers in 6E 12.5 format. All the trailing zeroes will be printed. To interrupt the printing of any block of numbers, type ↑T. The program then reads the next column vector and begins printing it in a new block of numbers. Terminate the program by typing ↑C. The program will not terminate itself at the end of the figure field.

Operational Instructions - RETRIV

Program name RETRIV

Function To retrieve a figure field from a named file

Written by D.S. Connelly

1. Introduction

RETRIV performs the reverse operation of the program ARCHIV. It retrieves a figure field placed in a file on an archive tape and writes it to a non-file-oriented tape. No previous information is retained on the output tape. Like ARCHIV it will process only files containing standard ECU figure fields. It is not a general purpose retrieval program.

2. Program setup

Mount the output MAGTAPE on unit 6, ring in place. Mount the archive (file-oriented) MAGTAPE on unit 5 with the ring removed. The DISC must be operating with DISC1 file-oriented, write-enabled and containing at least 200 free blocks.

Mount the contouring operations DECTAPE on unit 2. Then initiate the DEC MONITOR and type E RETRIV to fetch the program.

The program then requests the name of the file that is to be fetched.

Q1 GIVE NAME OF INPUT FILE (6 LETTERS + BIN)
R1 CHALK3BIN

Type the name of the file you wish to retrieve. Do not leave a space between the sixth and seventh characters of this name.

If this file is not present on the archive tape, Q1 will be repeated. Check that the name you have typed is correct and includes the characters BIN.

When the figure field has been written to the output tape, the program will begin again with the request for the input file name. If you wish to retrieve another figure field, mount a new output tape and supply the requested name. If you are finished, type ↑C to return to the MONITOR.

Operational Instructions - SAMPLE

Program name SAMPLE

Function To accumulate totals of points and line lengths for either individual features or features grouped by codes.

Written by D. Vicars

1. Introduction

The program processes each file individually. The files can either be selected one at a time in any sequence or the program can be made to process every file on the tape in sequence. The output is to paper tape (for subsequent printing on the off-line teletype).

2. Program setup

The input tape is mounted on MT5 (write disabled). The program first types its name, mark number and date of issue, followed by

Q1 DO YOU WANT TO PROCESS ALL THE TAPE
R1 NO

If the reply is NO then the program asks

Q2 ENTER FILE NUMBER
R2 3.

Q3 DO YOU WANT RESULTS FOR INDIVIDUAL FEATURES
R3 NO

The reply NO gives results summed over code, otherwise the results are listed for each individual feature.

Q4 ENTER SAMPLE CODE NO.
R4 3.

This is the position number of the code in the feature code list to be used in the sampling.

If the reply to Q1 is YES then the program will process every file on the tape, typing the elapsed time after each file. Otherwise the program will process only the indicated file, type the elapsed time, and then ask -

Q5 DO YOU WANT ANOTHER FILE
R5 YES

The reply "YES" causes the program to loop to Q2. Otherwise the program will rewind the input tape and stop.

3. Error messages

IOPS4 Either the tape bunch has run out of paper tape or the input magnetic tape is not online. Make the appropriate device ready and type CTRL and R to continue.

Operational Instructions - SLOPES, SHADOW, WAVER

Program names SLOPES, WAVER, SHADOW

Function Compute a figure field from data points

Written by D.S. Connelly

1. Introduction

The method of operation of each of these programs is roughly the same. Each of them reads an X, Y, Z list and produces a figure field of interpolated heights. The characteristics of the different routines are as follows:

WAVER - The routine interpolates the grid nodes by simply taking the weighted average of the closest 4 to 10 points. The weighting function is very large for data points close to a grid node. Smoothing is accomplished by limiting the size of the weighting function. WAVER is the fastest and crudest of these routines.

SHADOW - This routine is similar to WAVER, but it allows closer data points to shadow or dominate data points lying behind them. This improves the accuracy of the representation of steep slopes. The closest 6 to 10 points are used to reduce the chances that all but one of the points are shadowed.

SLOPES - This is similar to WAVER, but it includes a calculation based on the computed gradients at the data points. This eliminates the undesirable flattening of the surface at the data points.

2. Restrictions

The more sophisticated interpolation schemes run much slower than the simpler ones. A fifty by fifty figure field will take about ten minutes on the PDP-9 when WAVER is used. Time is one of the primary limitations on the use of these routines. The exact timings, though, are dependent on the distribution of the data points and cannot be predicted in advance.

All of these routines are capable of producing 201 by 201 figure fields if allowed to run long enough. It is doubtful if the figure field size per se will ever be a limitation on the use of these programs.

Besides running time, the chief restriction imposed on these routines is the storage space allowed for the X, Y, Z data point list. These routines have been designed to accommodate the largest feasible number of data points on the PDP-9. In some cases execution speed has been traded for storage space. As currently constituted the XYZ list length restrictions are as follows:

SLOPES	-	1350 data points
WAVER	-	3000 data points
SHADOW	-	3000 data points

It should be stressed that the number of data points handled is limited only by machine size. Both SLOPES and WAVER can process 5000 points on a CDC 6600.

3. Program setup

All of these routines should be fetched from the figure field operations tape. Mount this tape on DECTAPE drive 2. Mount the DECTAPE which contains the correct X, Y, Z data point file on DECTAPE drive 1. Neither of these drives should be write-enabled. Mount a clean IBM tape on MAGTAPE drive 5 with the ring in place. When all of these tapes are ready, initiate the DEC MONITOR and type "E" and a space followed by the program name.

After an introductory header which reminds the operator to mount the input and output tapes, the program requests operator activity.

1. Request: MOUNT PAPER TAPE
   PAUSE ØØØØØØ

   Response: Load the paper tape containing the correct figure field parameters. This is the tape generated by program FIGURE. When this tape is ready, type ↑P.

The program now reads the figure field parameter tape. If the number of data points in the file is less than four or greater than the maximum number allowed by this routine, the program gives the following advice:

2. Advice: ERROR**NUMBER OF DATA POINTS = (NUM)
   MAXIMUM = (MAXPTS)

   Response: This error is irrecoverable. Correct the tape and start again.

If the data point file is not on the DECTAPE on drive 1, the program notes this fact.

3. Advice: (FILENM) IS NOT ON DECTAPE
PAUSE ØØØØØØ
Response: Mount the correct DECTAPE on drive 1 and type ↑P. Note: in some versions of these programs, this fault will be noted by the message IOPS13. In those cases the program must be restarted.

Should any data point fall outside the mapping region the following message will be printed, but processing will not be interrupted.

4. Advice: POINT NO. (INUM) OUTSIDE REGION
Response: This message is advice only. To kill the job, type ↑C. No action is required if data points outside the region are acceptable. These points are treated just like all other points.

After reading in the complete list and ordering it, the program then calculates and writes the figure field to IBM tape. The figure field is written one column vector at a time. The length of the job may be estimated by noting the time between tape movements. Multiplying this time by the number of column vectors in the figure field gives the approximate duration of the remainder of the job.

When the figure field is complete, the elapsed time is printed and the output tape is rewound. The program then STOPs.

Note: Due to a mysterious bug in the DEC software, the tape rewind may cause the failure IOPSØØ. This causes the loss of the rewind and the time message only. The figure field is not damaged. If you are in doubt, verify the figure field with PIP. If the verification halts on an end of file mark as it should, the figure field may be assumed to be correct.

Operational Instructions - SPP

Program name SPP

Function To plot Gealux point symbols on film

Written by T. Lang

1. Introduction

Before using the Gealux to draw a map it may be necessary to investigate the appearance of various point symbols or combinations of point symbols to see if they are visually satisfactory. The Symbol Placement Program facilitates such an investigation. The user may divide a rectangular plotting area into a regular matrix with any number of rows and columns. He can then supply a data tape to plot any symbol at any position defined by a row and column intersection. Any number of symbols may be combined in one position.

If register marks are required to enable the results of separate drawings to be overprinted, the user should arrange for the data tape to specify suitable register symbols to be plotted at the corner points of the plotting matrix.

2. Program setup

Set up the Gealux, with its origin positioned near the bottom left corner of a rectangular piece of film (leaving a small margin).

The program starts by printing its title, and then asking a series of questions on the teletype:-

Q1 ENTER MAX X DIMENSION IN CMS
R1 10.0

Q2 ENTER NO. COLUMNS
R2 5.

Q3 ENTER MAX Y DIMENSION IN CMS
R3 10.0

Q4 ENTER NO. ROWS
R4 6.0

These questions establish the plotting matrix for the film. The above example would result in a matrix of plotting positions -

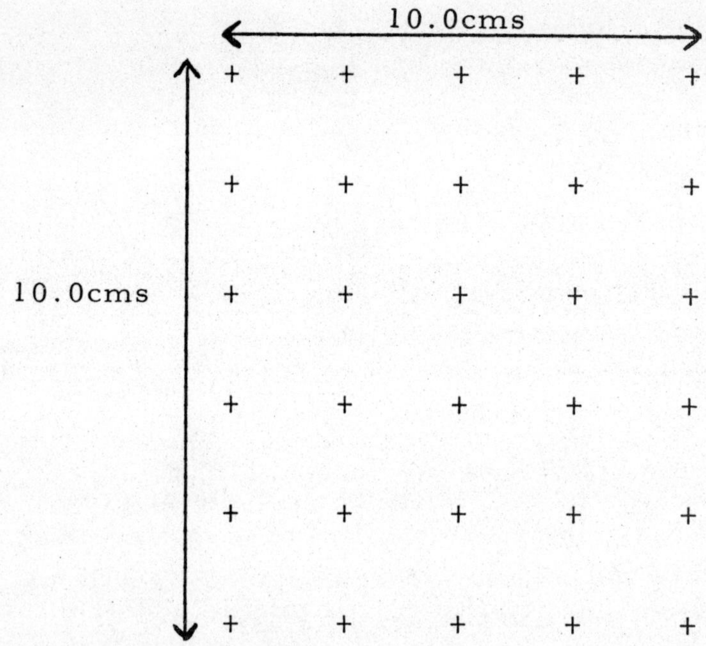

The program then outputs the message

LOAD PAPER TAPE READER
PAUSE

The data tape specifying the symbols to be plotted, and punched according to the format specified below in section 3, should be loaded in the paper tape reader. Type P to continue from the pause. The program will then read the data tape and plot the corresponding symbols.

When all the data has been processed, the program will output the message

END DATA REACHED

The program then returns to question Q1. At this point the user may

(a) terminate the program by typing ↑C on the teletype

(b) replace the exposed film by new film for a further test

(c) change the magnification but continue with the same piece of film to investigate the effect of a mixture of symbol sizes.

3. Data tape format

One line of data specifies one symbol to be plotted. The format for each line is

Row number - 2 digits. Rows are numbered from top to bottom, starting with row 1.

Column number - 2 digits. Columns are numbered from left to right, starting with column 1.

Symbol number (i.e. position on symbol disc) - 3 digits.

The end of data is denoted by specifying a row number of zero.

Example:-

```
_1_1_22        (where _ denotes a "space"
_2_1_21         character)
_3_1_28
_3_1_14
_0
```

Symbols may be superimposed by specifying more than one symbol with the same row and column numbers.

(N.B. Lines of data are terminated by Carriage Return and Line Feed characters. The first line must not be preceded by a Carriage Return, but may be preceded by a Line Feed.)

Operational Instructions - STG (MK 1B)

Program name	STG
Function	To convert a file on magnetic tape in standard internal ECU format to the standard format for exchange of cartographic data.
Written by	T. Lang
Revised by	D. Cranch

1. Introduction

The need for a standard format for exchange of cartographic data is set forth in "A standard format for the exchange of cartographic data on magnetic tape" by T. Lang, and in which the format is specified.

The present program reads a file from IBM tape in standard internal ECU format and prepares an output tape in Fortran BCD (DEC system). The output tape should be processed by IBMOUT to obtain a truly industry-compatible tape.

Messages are typed, followed by a pause, when certain errors are detected in the input data, i.e.

LINE POINT WITHOUT LINE

is typed whenever an internal line point, end of line, or end of loop is encountered without a corresponding preceding start of line.

START LINE IN MIDDLE OF LINE

is typed whenever a start of line is encountered following a start of line not having a corresponding end of line or end of loop.

Upon restarting after the pause, the line point or start of line, as the case may be, is not transferred to the output tape.

2. Program setup

The input ECU data tape should be loaded on MT5, write disabled, the output (exchange) tape should be on MT6, write enabled.

The program begins by typing its title mark number and date of issue. This is followed by typed loading instructions and a pause.

Upon restarting, the program types

Q1 GIVE INPUT FILE NUMBER
R1 93. for example

Q2a WRITE UP TO 10 LINES OF HEADER INFORMATION LINE 1
R2a ECU

Q2b LINE 2
R2b THIS IS AN EXAMPLE SHOWING HOW AFTER

Q2c LINE 3
R2c EACH LINE IS TYPED AND THE CARRIAGE

Q2d LINE 4
R2d RETURN KEY IS DEPRESSED, THE PROGRAM

Q2e LINE 5
R2e TYPES THE NUMBER OF THE NEXT

Q2f LINE 6
R2f LINE UP TO 72 CHARACTERS MAY BE

Q2g LINE 7
R2g TYPED ON EACH LINE

Q2h LINE 8
R2h

Q2i LINE 9
R2i

Q2j LINE 10
R2j

Following this, the program attempts to find the requested file. If unsuccessful the program types:

FILE 93 NOT FOUND

and loops back to Q1.

If the file is found, the program transfers the data to the output tape, changing the format, until either some faulty input data are discovered, in which case one of the two messages referred to in the introduction are typed out (the program will, upon being restarted, ignore the faulty data and pass to the next item), or until the end of the requested input file is reached.

If any feature is found to have more than 17 feature codes (the maximum for which the standard format caters) the following message is typed, followed by a pause:

WARNING ---- MORE THAN 17 FEATURE CODES

If the operator depresses CTRL P, the program will transfer the first 17 feature codes from the input tape to the output tape and continue.

When the end of the input file is reached, the program types:

Q3 ANY MORE FILES ON THIS TAPE
R3a YES - the program looks back to Q1
R3b NO - the program passes to Q4

Q4 ANY MORE INPUT TAPES
R4a YES - the program goes to Q5 and pauses.
R4b NO - the program types:
FINISHED

Q5 LOAD NEW INPUT TAPE ON MT5
Upon restarting after the pause, the program goes to Q1.

Operational Instructions - STNDFF

Program name	STNDFF
Function	To retrieve a figure field and write it in the standard data exchange format
Written by	D.S. Connelly

1. Introduction

STaNDard Figure Field fetches a figure field from an archive tape (similar to RETRIV) and writes it as characters in the form currently used for figure field data exchange (similar to FFOUT). Unlike FFOUT it permits a number of figure fields to be strung together on the output tape before the end-of-file marker is written.

2. Program setup

Mount the output MAGTAPE on unit 6, ring in place.
This tape must be positioned at beginning-of-tape manually by rewinding the tape once it is loaded. (REWIND switch) Mount the archive (file-oriented) MAGTAPE on unit 5 with the ring removed.
The DISC must be operating with DISC1 file-oriented, write-enabled and containing at least 200 free blocks.

Mount the contouring operations DECTAPE on unit 2. Then initiate the DEC MONITOR and type E STNDFF to fetch the program.

The program then requests the name of the file that is to be fetched.

Q1 GIVE NAME OF INPUT FILE (6 LETTERS + BIN)
R1 CHALK3BIN

Type the name of the file you wish to retrieve. Do not leave a space between the sixth and seventh characters of this name.

If the file is not present on the archive tape, Q1 will be repeated. Check that the name you have typed is correct and includes the characters BIN.

When the figure field has been written to the output tape, the program asks if more files are to be put on the same output tape.

Q2 DO YOU WANT TO PUT ANOTHER FILE ON
R2 YES.

Type YES to fetch another figure field and add it to the output tape without an intervening end-cf-file mark. Type NO to write an end-of-file mark on the output tape.

In either case the program begins again with a request for the input file name. If no more files are to be processed, type ↑C to return to the MONITOR.

Operational Instructions - STNDRD

Program name STNDRD

Function This program transforms data in the standard form for exchange of cartographic data to the standard internal ECU form for further processing

Written by D. Vicars

1. Introduction

The ECU has defined a standard magnetic tape format for the exchange of cartographic data. A tape supplied in this format could not be used directly on the ECU system because

(a) The DEC operating system requires tapes to have special markers and character sets which are peculiar to DEC machines, and could not be included in any internationally accepted standard.

(b) Because the exchange of data between computing systems is fraught with hazards and complexities, the more sophisticated features of the ECU internal system had to be omitted from the standard for data exchange.

The problems introduced by (a) are overcome by first processing the standard tape with the program IBM-IN, which adds the special markers and characters necessary to allow the tape to be read by a program using the normal PDP9 operating system.

Program STNDRD may then be used to overcome problem (b), and transform the data to the internal ECU format. Once the data has been transformed by STNDRD, it can then be processed by any of the other programs from the ECU repertoire e.g. drawing on the Geagraph, projection change, etc.

2. Program setup

The Input tape must be mounted on MT5, (write disabled) and the output tape on MT6 (write enabled).

The program will type its name and date of issue, and then ask

Q1 IS THIS THE FIRST FILE ON O/P TAPE
R1 YES

If the reply is NO question Q2 will be omitted, and the output tape will be spaced forward, to a point just beyond the last recorded information.

Q2 ENTER TAPE NUMBER
R2 MT 43.

The reply to Q2 must be MT or DT then the number terminated with a decimal point.

The following printouts and questions are repeated for every file on the input tape.

The input five header information records of 120 characters are typed first (each one on to 60 character lines) so that the user can answer the questions for the permanent file description data.

Q3 ENTER SCALING FACTOR
R3 Ø. 75

This factor will be applied to all the data points.

Q4 ENTER MT CONVERTION UNIT
R4 5Ø.

This is to convert the input co-ordinate values into the standard units of 0.02mm used in the internal ECU system. E.G. if the input co-ordinates were expressed in units of 1 millimetre, 50. would be the correct reply; if in units of 0.01 inches, 12.7 would be correct.

Q5 ENTER PERMANENT FILE DATA
PROJECT REF
R5 O.S.

This is an abbreviated reference (5 characters) for the project.

Q6 DESCRIPTION
R6 BIDEFORD AREA. ROADS ONLY.

Only the first 25 characters are retained.

Q7 JOB NUMBER
R7 23

This can be any 5 digit number. It is intended to be a numeric project reference.

Q8 SHEET REF
R8 SS42SW422Ø

This is a complete reference for the sheet contained in the output file. Only 10 characters are retained.

Q9 ENTER YOUR INITIALS
R9 D. V.

Only five characters are retained.

Q10 ENTER DATE AS DD-MM-YY
R10 Ø3-Ø2-7Ø

Q9 and Q10 are omitted on subsequent passes through the program. It is assumed that the data remains unchanged.

Q11 ENTER RUN DESCRIPTION
R11 INITIAL INPUT

The reply should be sufficient to distinguish the run, but it can be left blank.

If there are no corner points on the input tape then the program will ask question 12, otherwise it will carry on to complete processing the input file.

Q12 DO YOU WANT TO ENTER CORNER POINTS
R12 YES

If the reply is YES then this program repeats question 13 four times, for SW, NW, NE and SE corners.

Q13 SW CORNER
R13a Ø.
R13b Ø.

The numbers read in are transferred to the output tape in the same way as the numbers read from tape. If the reply to Q12 is NO then no corner points are output. (Practically every program which might be used subsequently to process the output file will require the presence of corner points. However, a sheet without corner points can still be drawn).

When the current input file has been processed, the program will test whether there are any further files on the input tape. If there are no further files, the program will rewind both the input and output tapes and then terminate. If there is a further file on the input tape, the program will loop back to print out the header information blocks for this file and then repeat from question Q3 onwards.

3. Error Messages

(i) IOPS4 A magnetic tape unit is not switched "on-line". Switch to "on-line" and type CTRL and R together to re-continue.

Operational Instructions - TIDY MK 1A

Program name	TIDY
Function	To reorder map sheets on tape and delete unwanted sheets
Written by	D. Cranch

1. Introduction

An input IBM tape containing digitized map sheets (files) in the sequence in which digitizing was carried out, and also containing unwanted files, is copied partly or wholly to an output IBM tape in a file number sequence entered via the teletype.

When copying is completed, TIDY will optionally copy more files from the input tape or files from one or more further input tapes.

The list of entered file numbers is checked for duplicates and an opportunity is given to correct the list if duplicates are found. This facility can be used to correct any erroneously entered file number, for, by intentionally re-entering the erroneous number, the correction can be made when the program reports the presence of duplicates.

2. Program setup

The input tape must be mounted on MT5 write disabled, the output tape on MT6 write enabled.

The program begins by typing its name, mark number and date of issue. This is followed by the questions:

Q1 ENTER DATE AS DD-MM-YY
R1 24-03-70

Q2 ENTER YOUR INITIALS
R2 ABC

The first five characters of R2 will be stored

Q3 ENTER RUN DESCRIPTION
R3 THIS IS AN EXAMPLE

Q4 IS THIS THE FIRST FILE ON O/P TAPE
R4 YES or NO

If the answer is NO, Q5 is skipped.

Q5 ENTER TAPE NUMBER
R5 MT73

The replies to questions Ql to Q5 inclusive will be written into the header block of each file transferred to the O/P tape. These questions are asked only once each run.

Q6 ENTER FILE NUMBERS. MAX. NUMBER OF ENTRIES IS 5Ø
R6 17.
12
-16
30.
99999.

Here the numbers of the files required to be copied to the O/P tape are entered in the desired order. Each number must be terminated by a dot. The program understands a negative file number as shorthand for all files from the preceding number to the negative number. Thus, in the example, file numbers 17, 12, 13, 14, 15, 16, and 30 will be transferred. The list is terminated by 99999.

The maximum number of entries is 5Ø. If the number is attained the program will type
LIST NOW FULL. NEXT QUESTION IN A FEW MINUTES

Whether entering is terminated in this manner or by 99999. the program will check the list of entered file numbers for duplicates. If none are found, the program transfers the entered files and skips to Q10, otherwise this message is typed:

FILE NUMBER 19 IS DUPLICATED.

Q7 RE-ENTER FILE NUMBER BEFORE FIRST FAULTY ONE
R7 123

Q8 CONTINUE ENTERING FILE NUMBERS

All the file numbers following 123, in this example, should now be re-entered. When the list is terminated, the program will type out all the entered file numbers, preceded by:

CORRECTED LIST IS

and followed by

Q9 IS LIST OK
R9 YES or NO

If the reply is NO the program will loop to Q7, otherwise the program transfers the entered files and proceeds:

Q10 ANY EXTRA FILES FROM THIS TAPE
R10 YES or NO

If the reply is YES the program loops to Q6, otherwise:

Q11 ANY MORE INPUT TAPES

If the reply is NO the run is terminated, otherwise the program rewinds the current input tape and types LOAD NEW INPUT FILE, then pauses until the new input file has been loaded and CNTRLP depressed, and loops to Q6.

Operational Instructions - TTP

Program name TTP

Function To add a termination marker to an incomplete data tape

Written by T. Lang

1. Introduction

A data tape normally consists of a number of separate sheets or files. The last sheet on the tape is followed by a special end marker. The end marker is used by the programming system to detect the point where the current data ends and new data can be added.

Under normal conditions, after a program has finished adding data to a tape it replaces the end marker in the correct position. However, it could be that whilst the program was still adding data to the tape,

(a) there was a hardware failure

(b) there was a software failure (not unknown, even in the ECU)

(c) the user deliberately abandoned the program

in which case the sheet currently being written to the tape will be left incomplete, with no end marker. The tape termination program can be used to remove the incomplete sheet, and replace it with a valid end marker.

2. Program setup

Load the tape to be terminated on MT5, before calling the program. The tape unit should be write enabled.

The program will ask

Q1 ENTER NO. OF SHEET TO BE CHOPPED OFF
R1 74.

The reply must specify the number of the incomplete sheet. The program will read through the tape to the start of the specified sheet, backspace the tape, and write an end marker. Finally the program will rewind the tape, and terminate with the message

STOP

3. Error messages

If the tape unit is now switched 'on-line' or not write enabled, the message will be typed

IOPS4

Switch the unit to the correct state, and type ↑R to continue.

It is very important that the user should specify the correct number for the incomplete sheet. If the number specified is too low, one or more valid sheets will be lost from the end of the tape. If the sheet number is too high, the program will attempt to read beyond the end of the incomplete sheet, with unpredictable results. There is no protection against either of these failures.